Smart Data Discovery Using SAS® Viya®

Powerful Techniques for Deeper Insights

Felix Liao

sas.com/books

Contents

Preface

Analytics is playing an increasingly strategic role in the ongoing digital transformation of organizations today. To succeed on your digital transformation journey, however, it is critical to enable analytics skills at all tiers of your organization and scale beyond the traditional data science team. It is through people with both strong business domain knowledge and analytics skills that you can often find the most valuable insights and make the biggest impact.

At SAS, we believe analytics can and should be for everyone and SAS Viya was built from the ground up to fulfill this vision of democratizing analytics. With a visual-based approach that supports the end-to-end analytics life cycle, SAS Viya supports the needs of traditional programmers as well as supporting a low-code and no-code approach to programming. By leveraging augmented analytics capabilities and machine learning based automation, we are making analytics easier and more accessible for everyone within an organization.

In this book, Felix Liao takes you on a tour of how SAS Viya empowers any user to uncover deeper insights using powerful analytics techniques. Felix reminds us that there is so much more to visualization today beyond just using traditional charts and graphs. Through step-by-step examples using real world data, the book guides the reader through how to apply statistical and machine learning techniques using a visual framework in order to answer complex business questions and extract valuable insights.

Shadi Shahin
Vice President, Product Strategy
SAS

About This Book

What Does This Book Cover?

This book focuses on how smart data discovery can empower everyone in an organization to leverage data in powerful ways and derive valuable insights. By leveraging the powerful visual interface of SAS Viya and more advanced analytics and machine learning-based techniques, the book demonstrates how to analyze business problems in new ways as well as derive actionable insights quickly and easily.

The main topics covered in this book includes the benefits of smart data discovery, the overall approach to smart data discovery, as well as how to apply specific smart data discovery techniques to solve business problems using SAS Viya.

This book does not cover how SAS Viya can be used for data discovery using programming techniques nor does it cover advanced modeling concepts such as pipeline modeling or model management.

Is This Book for You?

The intended audience of this book consists of business users and analysts who want to leverage data and analytics to drive actionable insights across multiple business functions.

It is for people who are familiar with traditional reporting or data visualization techniques but want to tap into the power of more advanced analytics and machine learning techniques in order to tackle more complex business questions.

What Are the Prerequisites for This Book?

While it would be helpful to have some familiarization with SAS Viya, there are no real prerequisites that are necessary in order to benefit from reading this book. This book introduces foundational knowledge that is needed in order to leverage more advanced statistical and machine learning techniques. Because it focuses on a visual approach to insight discovery and modeling, there are also no knowledge requirements around any programming languages.

What Should You Know about the Examples?

This book includes step-by step examples that you can follow to gain hands-on experience with SAS Viya. These examples use real data and demonstrate how specific analytics techniques can be applied in order to answer complex business questions.

Software Used to Develop the Book's Content

The examples used throughout the book leverage SAS Viya 3.5.

Example Code and Data

The examples in the book use the World Development Indicators data set published by the World Bank. You can access the data via the following link:

http://datatopics.worldbank.org/world-development-indicators/

We Want to Hear from You

SAS Press books are written *by* SAS Users *for* SAS Users. We welcome your participation in their development and your feedback on SAS Press books that you are using. Please visit sas.com/books to do the following:

- Sign up to review a book

- Recommend a topic

- Request information on how to become a SAS Press author

- Provide feedback on a book

Do you have questions about a SAS Press book that you are reading? Contact the author through saspress@sas.com or https://support.sas.com/author_feedback.

SAS has many resources to help you find answers and expand your knowledge. If you need additional help, see our list of resources: sas.com/books.

About The Author

Felix Liao is a manager within the customer advisory team at SAS and is also responsible for the analytics platform product portfolio for SAS Australia and New Zealand. He has over 15 years of experience working in the Australian and New Zealand analytics market. Felix was responsible for the regional launch of SAS Viya and was also responsible for the successful launch of SAS Visual Analytics in Australia and New Zealand in 2012. He is a regular speaker and blogger on the topic of analytics, data visualization, and machine learning. A computer engineer from his undergraduate study, Felix obtained his MBA in 2009 from Macquarie University, and he is also a SAS certified data scientist. His diverse background allows him to bring a wide set of views and perspectives, which are critical in modern analytics and machine learning projects and initiatives.

Learn more about this author by visiting his author page at http://support.sas.com/liao. There you can download free book excerpts, access example code and data, read the latest reviews, get updates, and more.

x

Acknowledgments

This book would not have been possible without the inspiration and support that many people provided me. I made many mistakes and leaned on many people for both their expertise and ongoing support. First of all, this book was inspired by someone who showed me the merits and joy of sharing your work and passion. Evan Stubbs who I had the pleasure of working with and inspired many people with his many popular SAS books really opened my eyes and planted the seeds for this book many years ago for which I am truly grateful.

As a first-time writer, my editor Lauree Shepard has been my constant rock in providing me tremendous support along the way. This book took a lot longer than both us thought initially, but Lauree never had a doubt. Her patience and guidance got me through many dark times during the writing process, for which I am truly thankful. I have only ever worked with one book editor, but I am not sure it can get any better than Lauree!

I also had help from many other SAS staff along the way who not only humbled me with their excellent analytics knowledge but also showed me how to be a better communicator and writer. People like Suneel Grover and Renato Luppi helped review early drafts of the book and set me on the right path for which I am thankful. I would like to especially thank Annelies Tjetjep and Sarah Gates, who guided me through the final editing and technical review process. They not only graciously offered significant amounts of their time, but they provided me with the honest and sometimes brutal feedback that I needed. Final technical review and editing is often a thankless task, and I want to acknowledge their significant contributions and thank them from the bottom of my heart.

Finally, I would like to thank my wife Elaine and my two kids Sophie and Lucas. Whilst they still don't quite understand what the book is about, they offered me unconditional love and support along the way. They understood the importance of Daddy's "Book Project" and gave me the time and space I needed at home. I could not have done it without their support. I love you all dearly.

Chapter 1: Why Smart Data Discovery?

Introduction

As information workers, our ability to leverage data and extract insights in order to make critical business decisions is fundamental to our success as individuals and the organizations that we work for. Regardless of whether you are an executive, departmental decision maker, or an analyst, the need to leverage data and analytical techniques effectively in order make business decisions is now pervasive throughout every part of an organization.

From the organization's perspective, the historical approach of relying solely on statisticians or decision support specialists to prepare and analyze data is no longer a workable approach. Organizations today must involve everyone in their analytics efforts – especially those closest to core business functions – to truly leverage data for maximum strategic and tactical advantage.

The good news for both us as individuals and the organizations that we work for is that tremendous advancements in computer hardware and software in recent years have allowed us to collect more data than ever before. Furthermore, with the addition of advanced analytics and machine learning capabilities, modern analytics tools are now easier to use and have never been more powerful. These shifts have made data more accessible and true self-service analytics a reality today.

One area of analytics that has made a significant impact in recent years is self-service data visualization. These easy-to-use data visualization and exploration tools enable any information workers today to assemble data rapidly, explore hypotheses visually, and find new insights quickly. Data visualization tools empower business users and accelerate the process of insight discovery by reducing the need for statisticians, data modelers, or IT specialists. By shifting the process of insight discovery closer to the business and subject matter experts, it has enabled more timely and relevant insights to be discovered and acted upon.

> "The greatest value of a picture is when it forces us to notice what we never expected to see."
>
> – John Tukey

Not only have these new data visualization tools accelerated the process of insight discovery, they have also allowed business users to ask more complex and forward-looking questions. These powerful, visual-based data discovery tools have revolutionized the traditional business intelligence solution space and led the way in terms of self-service analytics. Never before has it been easier for individual users to explore and visualize data for powerful insight with such ease.

With growing awareness and understanding of advanced data visualizations techniques, business users are now increasingly asking more complex questions, conduct more forward-looking analysis and eager to move beyond basic charts and graphs for answers. Enter the era of smart data discovery and the rise of citizen data scientist. Smart data discovery extends beyond the realm of traditional charts and visualization techniques with embedded machine learning techniques and algorithms. This new, augmented approach to data discovery leverages new visualization frameworks and automated machine learning capabilities to empower a new generation of users often described as citizen data scientists. Smart data discovery enables deeper insight discovery and empowers these new citizen data scientists to conduct deeper investigation, ask forward-looking questions, and develop valuable predictive insights.

What Is a Citizen Data Scientist?

Gartner defines a citizen data scientist as "a person who creates or generates models that leverage predictive or prescriptive analytics, but whose primary job function is outside of the field of statistics and analytics."

According to Gartner and other industry analysts, citizen data scientists are "power users" who can perform both simple and moderately sophisticated analytical tasks that would previously have required more expertise. They provide a complementary role to expert data scientists.

While smart data discovery holds tremendous promise, a new approach is needed in order to fully release its potential. At the core of this new approach is the recognition that citizen data scientists will need to be equipped with a new set of knowledge and skills, including the following:

1. **Business Domain Knowledge** – Smart data discovery needs to be built on a deeper understanding of the relevant business context and problem domains. Smarter insights can only come from asking more relevant and intelligent questions.
2. **Analytics Techniques** – A high level of familiarity with various analytical techniques and principles is required. While a PhD in Statistics is not necessary, a sound understanding of fundamental statistics and machine learning principles and techniques will be needed.
3. **Product Know-How** – Finally, it is about having access to the right tools and the necessary skills to bring it all together in a timely manner.

As depicted in Figure 1.1 below, these three skill sets in isolation are valuable, but when combined to solve a specific business problem, a new level of analytics and insight can be achieved – in this case, the sum is greater than its parts.

Figure 1.1: Smart Data Discovery Requirements

This book will help you navigate these intersecting knowledge domains and empower you to ask more complex questions by illustrating the key components of a smart data discovery process. We will highlight fundamental statistical concepts and how to leverage the relevant features. Most importantly, we will also be using real examples and applications to bring these concepts together.

SAS Viya, the latest evolution of the SAS platform will be used to demonstrate these examples throughout this book. An introduction to relevant features and functionalities that are needed in a smart data discovery process will be provided, followed by an explanation on how to leverage and interpret the various charts and outputs from SAS Viya.

Smart data discovery has the potential to shift an organization's overall analytic maturity, accelerate its analytical efforts, and create a much bigger analytics workforce. From an individual perspective, it has the potential to transform the way you view data, conduct data discovery processes, and think about how complex business problems can be solved. In many ways, we are just at the start of this revolution, and I am hopeful that this book will help you and your organization lead the way in terms of realizing the true potential of data and analytics.

Why Smart Data Discovery Now?

Traditional Business Intelligence (BI) and data visualization tools do a great job of slicing and dicing data in order to help answer questions such as what happened and what is happening. These tools can also provide valuable dashboards and reports for the purpose of insights sharing and communication. However, they cannot easily identify correlating factors or help predict future outcomes. These outcomes might include: which customers will respond to a promotion? Who will churn to a competitor? And when will a piece of equipment fail? In a modern environment where businesses are no longer content with simply analyzing data from the past, but instead would also like to gain insights into the future, a new approach is clearly needed.

Smart data discovery extends the traditional data visualization paradigm by marrying intuitive visualizations with predictive analytics techniques and machine learning algorithms. As illustrated in Figure 1.2, the use of these more advanced techniques changes the nature of analysis from reactive to proactive. It changes the perspective from backward-looking to forward-looking, and it empowers the analytics professionals to dig deeper, investigate root causes, and predict future outcomes.

Figure 1.2: From Reactive to Forward-Looking and Proactive

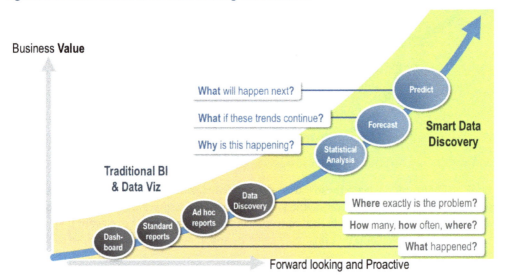

What is not widely known is that data visualization has always been a fundamental tool for expert data scientists. Visualization techniques are often used by expert data scientists for correlation analysis, model feature selection, and testing different hypotheses quickly. Techniques range from the use of box plots and histograms for general exploration to the use of decision tree and scatter plots for feature selection. These exploration steps often act as the precursor to more complex predictive modeling techniques. Smart data discovery extends these analytical techniques to a much broader audience using an intuitive and visual-oriented approach that does not require complex programming or deep statistical knowledge.

From an organization's perspective, smart data discovery has the potential to minimize the challenges associated with resourcing and staffing that most organizations face today. In an environment where there is still severe shortage of experienced, expert data scientists, smart data discovery not only has the potential to empower more users to leverage advanced machine learning techniques, it can also reduce the friction between the expert data scientist and the broader analytics professional community.

Data science is widely recognized as a highly collaborative process that require inputs from multiple teams and different personas. Expert data scientists are important and valuable, but they are only one part of the puzzle. Expert data scientists typically have a strong math and programming background and have a high-level understanding of the business process and functions. They normally do not have in-depth knowledge of the functional parts of a business versus someone who is directly involved in a particular line of business. Problems associated with marketing, fraud, customer service, and the supply chain all require deep and relevant domain knowledge in order to solve. This is where citizen data scientists fit in. They are typically used in lines of business and are intimately involved with core functional business areas. As a result, they tend to have a greater understanding and appreciation of the challenges being faced and how to solve them. This deep domain knowledge is critical in the success of smart data discovery processes. Armed with a new approach to data exploration and a powerful solution, smart data discovery can not only empower these citizen data scientists to solve more complex business problems, it can also bring business communities closer to the expert data science teams, which can only be a good thing.

Empowering and developing the citizen data scientist community will also benefit the expert data scientist community in many ways. As citizen data scientists improve their ability to ask more complex questions and test hypothesis more quickly, they can then communicate their findings with the expert data scientist community in a timelier manner and direct them to the more relevant, high-value problem domain areas. The expert data scientists will appreciate the deeper insights and analysis generated by the citizen data scientists, which will allow them to better prioritize and focus their time and efforts. This paradigm shift will lift an organization's overall analytical capabilities as well as create a more analytics-focused culture.

Who Is This Book For?

Whether it be basic descriptive analysis or sophisticated diagnostic and predictive analysis, the need to leverage data and analytical techniques in order to make important business decisions is everyone's business today. From that perspective, this book is really for everyone in every part of an organization. Having said that, this book takes a pragmatic approach and is targeted squarely at any analytics professional who need to bring in data, explore, prototype, and move the needle forward in terms of finding new insight and create insight value.

The SAS products that we will be using throughout this book are SAS Visual Analytics and SAS Visual Statistics. These two products form the foundation of SAS Viya and are targeted at non-programmers and business users. If you are an existing SAS Visual Analytics or SAS Visual Statistics user, you will benefit from the focus around advanced visualization and machine learning techniques covered in this book. We will be covering visualizations and features that are often unused or put into the "too hard" basket. The goal of this book is to help you realize the true potential and value of the tools that you and your organization have invested in.

On the other hand, if you are not familiar with the world of SAS Visual Analytics and SAS Visual Statistics, we will introduce the basic elements of each tool. You will find a dedicated chapter (Chapter 3) where we will provide you with a high-level overview of the key features and functionalities needed in a smart data discovery process. It is important to note that this book should not be treated as a training manual for SAS Visual Analytics or SAS Visual Statistics. We will not be covering every single feature or control offered by these tools. If that is what you want, there are many excellent books and training resources that cover these two tools in lot more depth, especially around the reporting capabilities that are really not the focus of this book.

While SAS Visual Analytics and SAS Visual Statistics offer powerful programming capabilities and user interfaces targeted at the programmers, those will also not be the focus of this book. We will primarily be focusing on the graphical interface. Once again, you can find out more about these programming capabilities via standard product documentations and SAS training courses.

What if you are intimidated by math? If you are in this group, then I have some good news for you: in order to get started leveraging machine learning techniques and build basic predictive models, you need less math background and knowledge than you think (and almost certainly less math than you have been told that you need!). While the role of a citizen data scientist does require a sound understanding of statistics and machine learning principles, you generally do not need to understand the mathematical underpinnings used to construct these algorithms in the first place. Modern statistical and automated machine learning software such as SAS Visual Analytics and SAS Visual Statistics take care of much of the mathematics for you, enabling you to focus on interpreting the outputs of these advanced techniques instead.

The only thing you are required to bring is your relevant business domain knowledge. Someone who has the relevant domain knowledge and is always generating interesting business problems and hypotheses will benefit greatly from this book. Ultimately, if you believe in the power of data and analytics and are curious as to how they can help you answer some of your most difficult questions, then this book is for you. I am hopeful that the combination of relevant foundational knowledge and practical advice in the following chapters will help unleash the true citizen data scientist in you.

Chapter Overview

While it might be tempting to jump into the chapter on prediction using decision trees to tackle your current business requirement, this book is written in such a way that it builds on the knowledge developed in prior chapters. The recommended way to read this book is to start from the beginning. Once you have finished the book, you can then use each chapter as a reference guide for specific challenges.

This book is loosely arranged into three main parts. The first part (Chapters 1 and 2) explains the what and why of smart data discovery and the role it plays for a citizen data scientist.

The second part (Chapters 3 and 4) provides foundational knowledge on the tools and data needed in a smart data discovery process. Chapter 3 introduces the relevant user interface components as well as key capabilities of SAS Visual Analytics and SAS Visual Statistics. Chapter 4 highlights the role data plays in a smart data discovery process and how to manage it effectively.

The third part (Chapter 5 through Chapter 10) dives into specific techniques as well as how these techniques can be applied to solve real business problems and extract valuable insights. While we do not delve into complex mathematical equations, various foundational statistical concepts are introduced throughout these chapters to support the examples and technique used. Chapter 10 concludes the book, but it introduces areas that you can explore further in order to build on the knowledge you have gained.

Chapter 2: The Role of The Citizen Data Scientist

The Rise of the Citizen Data Scientist

Despite the excitement and promise of data science in recent years, various industry surveys have revealed that the majority of organizations still have a shortage of data scientists today. The reality of most organizations that are trying to embrace analytics and data science is that it is extremely difficult to staff an adequate number of skilled data scientists simply due to the tremendous demand for them. Worse still, this trend is unlikely to change in the near future.

Given the scarcity of expert data scientists in the market and the ever-increasing demand for their skills, organizations are looking for an alternative approach and looking internally to accelerate their analytics efforts. As organizations look to uplift the skill level of existing staff, many are realizing that they already have people operating in a data scientist type of role and capacity, even if those employees don't realize it themselves or don't have that specific title. These people are often domain specialists in individual business departments or lines of business. They are often called "super users" or "data crunchers" and have been trying to address specific needs using different tools and approaches. They often have intimate knowledge of where the business problems or opportunities lie and where the relevant data resides, which is not always the case with the traditional data scientist. The key to success for organizations is to look internally and identify and develop these users by uplifting their skills. Enabling these users with a more robust framework and solutions set can help them reach the next level of effectiveness and productivity.

The term "Citizen Data Scientist" was first coined and officially defined by the analyst firm Gartner (see https://blogs.gartner.com/carlie-idoine/2018/05/13/citizen-data-scientists-and-why-they-matter/). Gartner defines a citizen data scientist as "a person who creates or generates models that use advanced diagnostic analytics or predictive and prescriptive capabilities, but whose primary job function is outside the field of statistics and analytics." In other words, citizen data scientists are business users or analysts who are in specific lines of business and are able to leverage data and advanced analytics techniques to solve complex business problems.

Individuals with solid business domain knowledge, familiarity with data challenges, and the willingness to embrace new methods of analysis are ideal candidates to fill the role of citizen data scientists. Existing domain knowledge in marketing, sales, or operations can often be leveraged effectively as long as it is supported by sound knowledge of statistics and machine learning principles. By uplifting the skill sets of existing employees, organizations can change the traditional paradigm around data science and machine learning. Instead of bringing the vast number of problems to the limited number of expert data scientists, we can bring the data science approaches and techniques to more business analysts who understand the business problems intimately.

Bring an Interesting Question

In analytics, success is often dependent on finding the right questions. Coming up with a relevant and high-value business question is critical because it helps you frame your analytical approach and empowers you to come up with innovative answers and solutions. It is also something that is uniquely suited to the role of a citizen data scientist. The business contextual and domain knowledge of the citizen data scientist is vital in working with the relevant business stakeholders and identifying high value business questions.

> "If you do not know how to ask the right question, you discover nothing."
>
> – W. Edwards Deming

Depending on the specific industry and the strategic focus of the organization, high-value, relevant questions posed by the citizen data scientists might include:

- Which types of customers have churned to a competitor?
- Why did these customers churn to a competitor?
- What factors contribute to high customer satisfaction?
- Who is likely to respond to a new marketing campaign?
- What is the best way to target and communicate with these customers?

These questions are just within the domain of customer insights. With the massive amount of data now available to us, we can ask more complex and probing questions across all areas of a business and expect to come up with a reasonable answer. Having said that, it can also be a double-edged sword. As the vast array of possible questions and paths of analysis open up to us, it becomes increasingly important to make sure that the questions being asked align well with the overall strategic goals and objectives of the organization. For example, the list of questions posed above would be appropriate for an organization with a focused strategy around customer centricity. On the other hand, a government department with a mandate around safer roads and reducing road fatalities will have a completely different set of questions. While it sounds simple and trivial, finding and focusing your efforts on the right questions can often be challenging before you even get to any data.

By choosing questions that align well with the overall objectives of the organization, it not only enables you to focus your time and resources more effectively, it also makes it easier to highlight and demonstrate the value provided through these analytical efforts.

Regardless of the area of focus, an insightful question should also enable you to eventually turn the analysis into actionable insights. When this is done effectively with strong alignment with the strategic objective of an organization, tremendous value can be unleashed through the analytical efforts. Extending from the example questions above and using the same types of customer data can lead us to extend the analysis and investigate the following questions:

- Which existing customers are likely to churn in the future?
- What is the best way to reach these customers?
- How can we increase the satisfaction of these customers to avoid them from churning to competitors?

By shifting the perspective of questioning into a forward-looking one, we start to uncover actionable insights that provide predictive value to an organization. If we can reasonably predict which one of our customers is likely to churn to a competitor and prevent that from happening, it leads to a clear business outcome and significant value for any organization. Being able to answer and explore these questions shifts the paradigm from one of being largely reactive and exploratory to one of being empowered to take proactive, valuable actions based on new, powerful insights.

Machine learning techniques make it possible to dig into these more complex, forward-looking questions and analysis as we will explore throughout the rest of this book. Whether it be reviewing historical purchase patterns, understanding correlating factors around customer satisfaction, or building a predictive decision tree churn model, smart data discovery allows us to tackle more complex questions using more advanced techniques in an intuitive and agile fashion.

Accelerate the Analytics Life Cycle

While there have been advancements in various specific analytics techniques over the years, the fundamental process of turning data into actionable insights has not really changed. The analytics life cycle is the general framework that organizations rely on to structure and staff their analytics teams and resources.

The analytics life cycle in its simplest form consists of three distinct steps (as shown in figure 2.1) that are sequential in nature but also form an iterative loop.

1. Sourcing and preparing data
2. Data exploration and model building
3. The insights generated in step 2 are then shared or deployed and embedded into business processes or applications in the final step.

As the data changes and the models decay over time, the process typically then starts again and forms an iterative loop.

Figure 2.1: Analytics Life Cycle

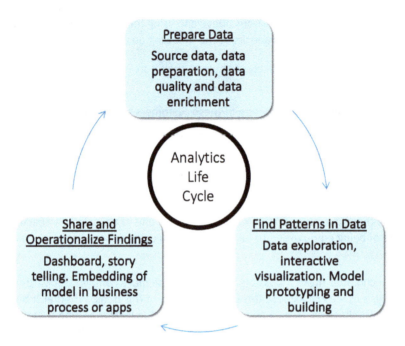

Depending on the types of data, specific business context, and the types of insights needed, the analytics life cycle can often take days, weeks, or sometimes months to complete. The ultimate goal of the analytics life cycle is to turn data into value in the shortest possible of time. Not only does this mean reducing the overall cost of analytics projects, it also improves key business metrics such as time-to-market and the overall competitiveness of an organization.

The agility and scale required around analytical efforts today have forced organizations to look for ways to improve upon it. Beyond the resourcing challenges associated with staffing traditional data scientists, data scientists can also be a bottleneck when it comes to iterating through the analytics life cycle quickly. Having only data scientists perform the function of pattern identification and model development in isolation can often create unnecessary friction in the analytics life cycle due to the potential for poor business alignment and lack of broader collaboration.

So how can citizen data scientists play a greater role and accelerate the analytics life cycle?

- Citizen data scientists can bridge the gap between the expert data scientist and the business analysts, especially during the data discovery phase. Having citizen data scientists involved during this phase means more patterns can be uncovered quicker and more relevant model prototypes built without the involvement of expert data scientists.

- Citizen data scientists can accelerate the path to value by filtering out poor analytics use cases and ideas early in the analytics life cycle. This means more relevant, high-value problems can get into the hands of expert data scientists quicker.
- Furthermore, citizen data scientists can often provide the expert data scientist with a validated analysis or early model prototype to explore, and this ultimately results in faster times to insight and action.

Communicate and Collaborate

One of the most underrated aspects of a citizen data scientist's job is communication and collaboration with key stakeholders. Just like all projects, the ability to demonstrate the value of your analysis and work is often key to ongoing support and funding. Moreover, lack of effective ongoing communication between the citizen data scientists and the other key stakeholders may lead the analysis down a path that is of little interest or use to the business, no matter how correct the analysis may be. Data and analytics alone will often not be enough to influence, gain trust, or effect change. It is often through the process of effective communications and powerful storytelling that you gain trust and inspire action.

> "Things get done only if the data we gather can inform and inspire those in a position to make a difference."
>
> – Mike Schmoker

It is during the process of communication and storytelling that the visual aspects of smart data discovery really shine over a programmatic or code-oriented approach. The interactive nature of visualizations helps to create an environment where the various stakeholders can try different scenarios and test different hypothesis on the fly together. Some of the ways that you can communicate your findings effectively that we will cover throughout the rest of this book include:

- Appropriate charts and visualizations
- Interactive filters and linked selection
- Playable dashboards
- Storyboard techniques

Furthermore, the use of interactive machine learning models during a smart data discovery provides a unique way to communicate and collaborate around complex business problems. This is especially the case when an interpretable model such as a decision tree model is used (Chapter 8). Not only does it allow the citizen data scientist to collaboratively explore and interpret the model in terms of relevant underlying factors, it also often can highlight relevant actions that can rectify the specific business problem or challenge.

For example, an interactive visualization using a classification tree algorithm to predict customer churn flag can be a powerful tool to analyze the drivers for customer churn and identify possible remediation actions. When combined with other traditional visualization objects such as filters and linked selections (Chapter 3), a citizen data scientist can effectively break down the scenarios and predictive model into different geographical regions or customer types. This facilitates deeper analysis as well as collaboration with other stakeholders in the area of customer insight. We will be expanding and discussing these technique and approaches in later chapters.

The late academic statistician and physician Hans Rosling was a great example of demonstrating the power of effective communication and storytelling using data and visualization. Hans tackled a number of global issues such as poverty and health issues in developing countries on the global stage. His influence in the world was built on his ability to communicate powerful messages through data visualizations and storytelling. His TED talks on the various world development topics are legendary and have raised awareness of these topics in powerful and meaningful ways. See his TED talk at
https://www.ted.com/talks/hans_rosling_the_best_stats_you_ve_ever_seen for an example of powerful storytelling using various visualization techniques.

I hope you are inspired and energized by what you can achieve and the impact that you can have as a citizen data scientist. Now it's time to get our hands dirty by tackling our very own challenging questions and telling our own powerful stories!

Chapter 3: SAS Visual Analytics Overview

Introduction

> "Man is a tool-using animal. Without tools he is nothing, with tools he is all."
>
> – Thomas Carlyle

Having the right tool and being able to leverage it effectively is critical to unleashing the value of a smart data discovery process. An effective tool will not only empower the user to ask complex, forward-looking questions and drive powerful analysis, it will also help teams to communicate and collaborate better throughout the analytics life cycle.

SAS Viya was built from the ground up as a comprehensive, fully integrated analytics platform that supports the complete analytics life cycle. It enables everyone – data scientists, business analysts, developers, and executives alike – to collaborate effectively and realize innovative results faster. SAS Viya accelerates the path to insights and value by supporting the smart data discovery process, advanced data modeling as well as the model management and deployment process. This book focuses on how specific SAS Viya components (SAS Visual Analytics and SAS Visual Statistics) can be leveraged to drive the smart data discovery process and accelerate the early parts of the analytics life cycle.

SAS Visual Analytics provides the primary graphical user interface (GUI) framework that the user interacts with when working in SAS Viya. Furthermore, SAS Visual Statistics provides more advanced analytics functionalities on top of SAS Visual Analytics and enables you to develop and prototype predictive machine learning models rapidly and easily as illustrated in Figure 3.1. SAS Visual Statistics provides users with access to additional analytical techniques such as logistic regression, linear regression and clustering, which we will cover in later chapters.

Figure 3.1: Visual Analytics and Visual Statistics Capabilities

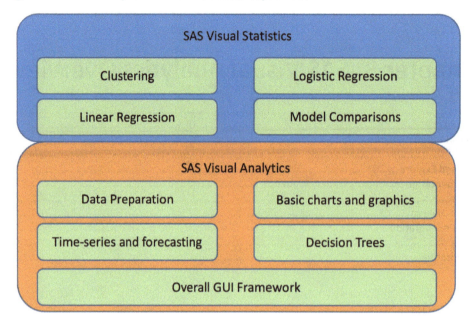

As well as providing the overall GUI framework, some of the other key functionalities provided by SAS Visual Analytics include:

- Data profiling and preparation
- Interactive reports and visualization
- Time series analysis and forecasting
- Predictive model prototyping (Decision Trees)

From the user's perspective, the GUI framework of SAS Visual Analytics provides a set of foundational features and capabilities that are common across different usage scenarios. These include important elements of the user interface as well as useful features that enable you to build dynamic and complex user interactions. This chapter will introduce and equip you with the foundational knowledge needed to leverage these key features. This foundational knowledge will not only help you get the most of SAS Visual Analytics, they are also necessary in order to tackle more complex analysis as well as build more dynamic and interactive reports and visualizations.

The User Interface

SAS Visual Analytics provides the foundational GUI framework and is the canvas for you to design reports, explore data, and build interactive, predictive models. As you select a new report or open an existing report, you are first presented with the main SAS Visual Analytics user interface as shown in Figure 3.2.

The main user interface is broken into four main areas: the left pane (area ❷ in Figure 3.2), the right pane (area ❸ in Figure 3.2), and the main canvas (area ❹ in Figure 3.2). The left pane enables you to work with data, objects, and the report outline. The right pane enables you to work with details about the selected chart or object. The menu bar (area ❶ in Figure 3.2) displays the report name as well as enabling you to do undo, redo, save, and access other menu options via the More button. We will be referring to these panes and menu options throughout the book, so it is important that you know how to find and locate these controls.

Figure 3.2: Main Visual Analytics User Interface

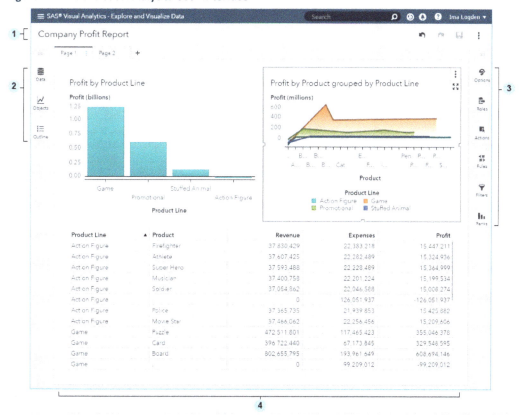

Using the Left Pane and the Right Pane

When working within the main SAS Visual Analytics user interface, a good tip to remember is that in general, data-related tasks are located on the left pane and presentation-related tasks are located on the right pane.

Panes in Visual Analytics

The Data pane (shown in Figure 3.3) on the left-hand side enables you to quickly add and select the data source, manage all the data items, and create custom data items. For categorical data items, the cardinality value is also displayed automatically to guide you on the right data items to use.

Figure 3.3: Data Pane

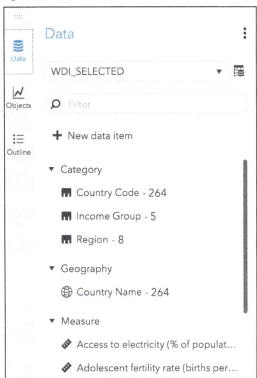

The Objects pane (shown in Figure 3.4) is where you choose and leverage the various visualizations and analytics objects. They are grouped by object type as well as specific SAS Viya component with all the Visual Statistics objects grouped under the Visual Statistics group as shown in Figure 3.4.

Figure 3.4: Objects Pane

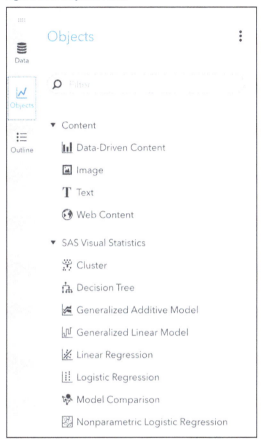

The Data pane and Objects pane are referred to and used throughout the book because they play the most import roles in a data discovery process. On the right-hand side of the canvas, the Options pane (as shown in Figure 3.5) lists the options for the currently selected report, page, or object. Depending on the selected object, the configuration options available in the Options pane will vary.

Figure 3.5: Options Pane

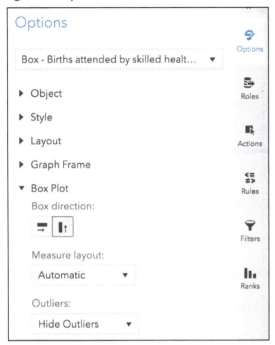

The Roles pane is one of the simpler panes to understand because it simply enables you to assign the data items needed for the selected visualization. The required data role type and number of data items will vary depending on the selected visualization as we will see throughout the book.

The Actions pane enables you to create interactions between objects such as Filter or Linked Selection, which are discussed in more detail in later parts of this chapter.

The Rules, Filters, and Ranks panes are mainly used to develop dynamic report behaviors and complex dashboard interactions and will not be covered extensively in this book as they are mainly used to build enterprise reporting and dashboard solutions.

Canvas

The canvas is the main workspace for building visualizations and predictive models and is where the magic happens. The canvas is also where you can leverage machine learning techniques and develop predictive model through the process of visual modeling, which we will cover in Chapters 7, 8, and 9. At its most basic level, the data discovery process is really just about dragging data items onto the canvas, assessing and evaluating the results, then changing the options and/or the types of objects used in order to get to the insights and the answers that you are looking for. The tool seems easy and intuitive because it was designed to be. Having said that, understanding and being able to leverage some of the more advanced elements of the GUI framework will help you to be more productive and undertake more complex analysis.

Experiment and Explore

Experimentation and exploration are at the heart of an effective smart data discovery process. SAS Visual Analytics was built with that in mind and offers specific features and functionalities to support ad hoc user experimentation. We will highlight a number of these functionalities that encourage you to experiment with new visualizations and try new data items.

Experimentation

When it comes to data exploration, analysts and data scientists often like to try different approaches quickly and fail fast so that they can quickly move on to a different approaches or hypothesis should the tried approach to be ineffective. Being able to undo and roll back to your previous step quickly means that there is little penalty for trying new data items or different visualization combinations. This can be easily achieved in SAS Visual Analytics by using the undo button in the menu bar. Furthermore, pressing and holding down on the mouse button with the cursor over the **Undo** button enables you to roll back to any of your previous steps (as shown in Figure 3.6) so that you have more control as to how many of your previous steps you want to keep. While extremely simple, undo is perhaps one of the most useful features and something you should really embrace and be familiar with as you try to use different data item combinations and experiment with different visualization types.

Figure 3.6: Press and Hold to Undo Multiple Steps

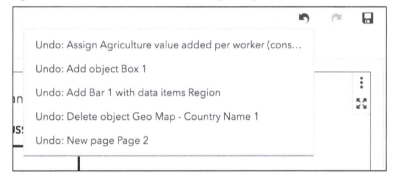

Another useful feature of SAS Visual Analytics that helps support the process of experimentation is auto-charting. Auto-charting automatically works out the best visualization type based on the type and combination of data items a user drags onto the canvas. As a user drags one or more data items onto the canvas, SAS Visual Analytics automatically selects a visualization type and displays the chart automatically without the user having to explicitly tell the tool which type of object he or she wants. The type of chart selected is based on both the type of data item selected (measure or categorical variable) as well as the combination of data items being dragged onto the canvas (one, two, or more data items). SAS Visual Analytics does not assume that you are an expert in data visualization, and the auto-charting capability enables you to simply focus on the business problem at hand. When in doubt, just drag the data items onto the canvas and see what happens. Remember: you can always override the automatically selected chart type, and there is always the Undo button!

Maximized View

Once you do find a visualization that is particularly insightful or useful, the Maximize feature enables you to focus on the single, selected visualization and reveals more details about the chart.

To maximize an object, click on the **Maximize** button from the object toolbar just below the More button. You can also use the shortcut **Alt+F11** to maximize the selected object. Maximizing a visualization object expands it to the full size of the canvas and fills the entire workspace. It eliminates all other visualizations on the same page and displays a details table underneath the visualization for the object. It is useful for diving into a specific chart and provides you with additional information and details. Figure 3.7 shows a maximized view of a box plot where the tabular data below the box plot displays key values associated with the box plot such as average, median, and standard deviation.

Figure 3.7: Maximized View with Tabular Data

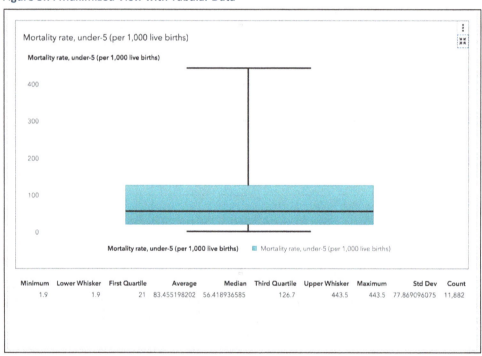

Minimum	Lower Whisker	First Quartile	Average	Median	Third Quartile	Upper Whisker	Maximum	Std Dev	Count
1.9	1.9	21	83.455198202	56.418936585	126.7	443.5	443.5	77.869096075	11,882

Speaking of experimentation, SAS Visual Analytics also enables you to easily transition from one type of visualization to a different type in order to gain a different perspective. Experimenting with different ways to visualize the same data is very much at the heart of a smart discovery process, and SAS Visualization makes it extremely easy and fast to switch between different types of charts. Clicking on the **More** button within the canvas provides you with the option to change the current visualization into a different type. Auto-charting is also at work when you try to transition into a different chart because SAS Visual Analytics will often highlight what it thinks is the best visualization for that data type. Figure 3.8 highlights an example where a histogram is a recommended chart as we try to transition away from a box plot chart.

Figure 3.8: Transition to Different Visualization Types

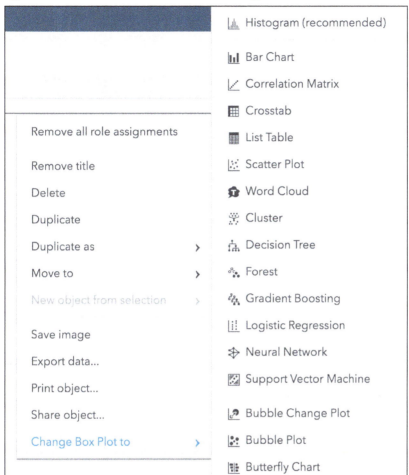

The built-in transition functionality eliminates the need to drag and drop the same variables or apply the same set of filters and reduces the amount of time needed to explore similar or related charts and visualizations. In other words, the software reduces the friction involved in experimentation and trying different approaches.

As well as changing the current visualization and data set into a different visualization type, you also have the option of duplicating the selected data set as a completely new visualization. The **Duplicate as** option made available when you press on the **More** button enables you to create a new chart using the same set of data items or filters. Looking at the same set of data side-by-side using different visualization types is another powerful way to interact with the data and uncover hidden insight. Let's look further at how we can leverage multiple charts more effectively by linking them together via actions.

Actions

The dynamic interactions between different charts and visualizations can often reveal powerful new insights. This is especially the case during complex analysis when you need to understand the complex relationships between multiple, interacting data items.

SAS Visual Analytics Actions enable users to understand data within a particular context across multiple charts. A SAS Visual Analytics action directs a report viewer's intention from one chart to another, and a typical application of this is the Filter action. When you select the **Filter** icon (the funnel icon in Figure 3.9), data in the linked chart is filtered based on the user's selection in the source chart. A SAS Visual Analytics action can also be in the form of **Linked Selection** (the brush icon in Figure 3.9), which highlights the affected data in the linked chart instead of filtering them. Both of these types of actions can be configured by using the Actions pane on the right-hand side of the canvas as shown in Figure 3.9. The Object Links section within the Actions Pane controls which objects you want to link to (the "Geo Map – Country Name 1" chart in Figure 3.9), and you can link to one or multiple objects on the same page.

Figure 3.9: Configuring Object Links Action

The Filter action is the most common type of linked action and provides an intuitive way to drive and interact with the charts and data on the same page. As you select data items on one chart, the linked charts on the same page will have their data filtered and focused around your selection.

Linked Selection on the other hand does not filter data on the linked object but highlights the linked data components based on the selection from the source object. Figure 3.10 shows an example of Linked Selection. In this example, the population bar chart (source) on the left-hand side is linked to the income group frequency bar chart (target) on the right-hand side as a Linked Selection. When we selected Sub-Saharan Africa as a region on the bar chart on the left-hand side, the original bar chart representing income groups on the right hand bar chart is kept, but you can see the linked portion of the bar chart represented as shaded regions, highlight proportion of the data that are from the Sub-Saharan region.

When we apply a business context and try to interpret the chart as shown in Figure 3.10 through the lens of linked selection, we can see that:

- Only a small proportion of all global countries are classified as low income countries in 2015 ("Low income" having the shortest bar).

- Most of these low-income countries (shaded region in the "Low Income" bar) are concentrated in the Sub-Saharan African region (selected on the left-hand side).

This is a great example of how linked selection can be used to highlight important insights across multiple, related contexts. By highlighting instead of filtering data in related charts, you can often get a more nuanced interpretation of the situation at hand.

Figure 3.10: Linked Selection Interaction

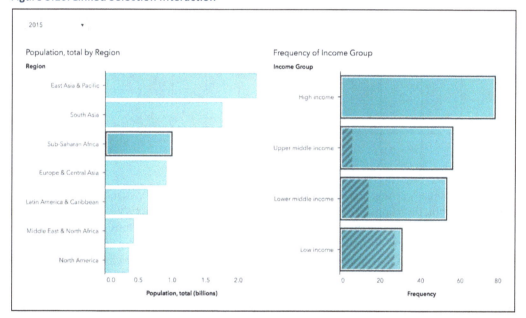

Actions Diagram

You can often end up with multiple actions (Filters or Linked Selections) on the same page, and it can be difficult to manage all the different actions via the Actions pane on the right. The Actions Diagram is an alternative to using the Actions pane when you have multiple linked actions. The Action Diagram feature makes it easier and quicker to view and edit these actions. The Action Diagram can be activated by clicking **View Diagram** in the **Actions** pane. As shown in Figure 3.11, the Actions Diagram enables you to link objects using either of the action types simply by dragging lines that link objects together. Not only does it provide you with a more complete view of all the linked relationships, it also makes it easier and quicker for you to modified specific links and the direction and types of actions across multiple pages.

Figure 3.11: Actions Diagram

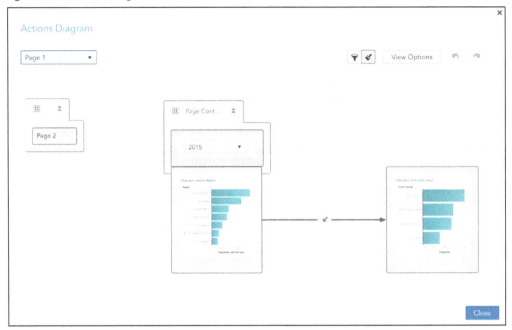

Automatic Actions

Instead of manually linking and defining individual linked actions, SAS Visual Analytics supports automatic actions to make the process even easier. In situations where you know you want to link all the objects, you can define this via the automatic actions configuration within the Actions pane as shown in Figure 3.12. Once automated one-way filters have been configured, all objects that you drag onto the canvas will automatically be configured to have one-way filters. This saves you the trouble of having to configure the individual linked actions manually yourself. Furthermore, you can also configure the automatic actions to be either two-way filters or linked selection depending on your specific situation and context. Automated actions are a great way to speed up your smart data discovery process.

Figure 3.12: Automatic Actions Configuration

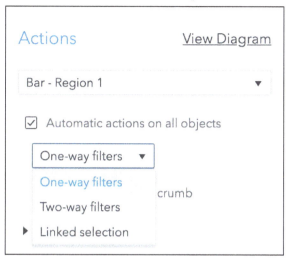

Sharing and Collaboration

As discussed in Chapter 2, communicating and sharing insights effectively is perhaps one of the most important aspects of smart data discovery. SAS Visual Analytics excels in that regard due to the fact that it is a reporting, data exploration, and modeling tool all rolled into one. This makes it uniquely suited to support a smart data discovery process that involves communication and sharing insights with one or many people. We will cover a number of reporting and sharing-related features here that help you to be a more effective communicator and collaborator.

Sharing

Once you have built a series of pages with charts and visualizations from your discovery process, SAS Visual Analytics makes it extremely easy for you to share them. Depending on the circumstances and the intended audience, you can share your reports and visualizations a number of different ways. The easiest and perhaps most effective way to share your insights is to share the complete report via email and links. Sharing via email or links allows the recipients to open the report in the Visual Analytics report viewer directly and provides the richest user experience and interactivity. When recipients click on the link, they are taken to the viewer mode where the reports are presented in a streamlined, viewer friendly format without all the editing tools. All of the standard interactions such as report prompt and interactive filters are available in the viewer mode.

To email the report, click the **More** button in the application menu bar, select **Share report**, then select **Email…** as shown in Figure 3.13. A new email message is generated and opened in your default email application with the subject line automatically populated with the report name. The link to the report is also automatically generated and embedded in the email body.

Figure 3.13: Sharing Report

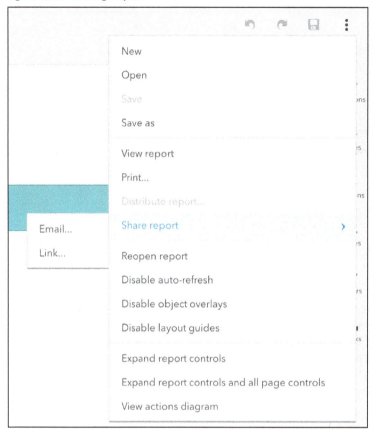

If you just want to share the link to the report via other message platform or collaboration tools, click the **More** button in the application menu bar, select **Share report,** then select **Link**.... The Generate Link window is displayed, which enables you to further customize the link as illustrated in Figure 3.14. The various controls enable you to customize permissions in terms of reporting printing and further sharing. You can also share the reports as static images (SVG files) if you choose the **Report Image** option as shown in Figure 3.14.

Figure 3.14: Link Generation Option for Individual Objects

As well as sharing links to the complete report, SAS Visual Analytics also enables you to share just a single object from your report. This is especially useful when you do not necessarily want to share the whole report with someone else but would like them to view and focus on one specific aspect of the report. This can be done by clicking the **More** button on the object toolbar instead of the application menu bar on top. With the menu button clicked, select **Share object**. A similar **Generate Link** window is displayed, which enables you to customize how you want to share the object.

In situations when your collaborators do not have access to SAS Visual Analytics, or in situations where a different, more static collaboration mechanism such as the Microsoft Office suite of tools is preferred or required, SAS Visual Analytics allows you to share your report by generating static images that can be embedded in other documents or sent directly. All individual charts and visualizations can be exported and shared as PNG image files in both editor mode and reader mode. To generate an image file, click the **More** button on the object toolbar, then select **Save image**.

Sharing reports can also be done in non-digital formats. In rare situations where a more physical medium is still needed, SAS Visual Analytics includes built-in support for printing the complete report or individual objects using the More button discussion earlier. The print function offers comprehensive printing options and controls such as page orientation, size, margin, cover page, and many others.

Collaborating

Collaborating around your discovery and insights should not stop once you have sent an email or link. The collaboration features within SAS Visual Analytics mean that you can also have ongoing discussions around the reports that you have just shared without relying on another third-party tool. This is done via the commenting capability available in the Visual Analytics viewer itself. This is another reason why sharing your report via direct link should be your default method. As shown in Figure 3.15, the commenting pane on the right-hand side in viewer mode enables you to create new topics as well as search for and participate in existing discussions.

Figure 3.15: Commenting in Viewer Mode

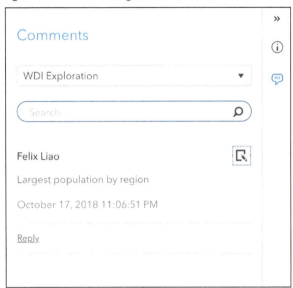

With the proliferation of mobile devices and a highly mobile workforce, insight sharing and collaboration do not always happen in front of a desktop or laptop computer today. SAS Visual Analytics was built to support mobile workers and collaboration on the go. Viewing of SAS Visual Analytics reports can be done using native apps supported on iOS, Android, and Windows Surface devices. These native apps are purpose-built and offer all the same browser-based viewer functionalities mentioned before as well powerful, device-specific capabilities that makes the user experience extremely rich and interactive.

The SAS Visual Analytics app includes advanced collaboration features such as **Present Screen** and **Annotation**, which enable you to annotate and share your screen on the fly. This is particularly useful when two mobile workers need to collaborate on a Visual Analytics report when away from the office. These mobile apps and unique features make the SAS Visual Analytics app uniquely suited for ad hoc collaboration and insight sharing on the go, and I encourage you to explore and try the app yourself.

Chapter 4: Data Preparation

Introduction

The process of importing, joining, cleaning, and transforming data is fundamental to the effectiveness of a smart data discovery process. However, data preparation is often the most time-consuming aspect of the data discovery process and can be the least desirable or interesting part of the overall analytics workflow. In this chapter, we will highlight and show how you can import and prepare your data easily and efficiently using SAS Viya.

It should be noted that the data preparation we are focusing on in this chapter is often described as "self-service data preparation" or "data wrangling." It is more ad hoc and less structured in nature and often exists as an extension of a traditional enterprise data integration process (think Extract-Transform-Load or ETL). In some situations, these types of data preparation can also be done completely independent of a traditional ETL process where data is not sourced from a traditional enterprise data warehouse or data mart (such as using a text file downloaded directly from the Internet). In any case, the data preparation covered in this chapter focuses on the ability for an analyst or data scientist to quickly access, import and transform relevant data so that insights can be discovered quickly and easily.

Regardless of the input data source types, having the right data in the right shape is fundamental to the data discovery process. This is particularly relevant when we cover the more advanced use cases using predicting modeling techniques in later chapters.

Importance of data management
It is commonly agreed that 80% of analytics project efforts are spent around data preparation and data wrangling. More importantly, the concept of "garbage in, garbage out," is even more relevant in the field of data discovery. A little extra effort and attention paid during the data preparation phase often pays massive dividends in terms of the accuracy of the insights and the time saved in downstream activities.

An agile and integrated data preparation process is key in helping citizen data scientists bringing data into a data discovery environment quickly and easily. It is about supporting the process of experimentation and prototyping with new data sources and different data variables. To that end, SAS Viya includes a number of built-in capabilities that supports the process of importing, profiling and transforming data. These integrated data preparation functionalities enable you to seamlessly transition from data preparation to data exploration to modeling.

Importing and Profiling Your Data

Built-in, deep, and mature data preparation functionalities make SAS Viya unique in the data exploration tools market. Not only can you easily import data from multiple sources, there are sophisticated data profiling and transformation functionalities built into the tool to support ad hoc, self-service, focused data preparation all within the same application framework.

Data Importing

The SAS Data Studio module is an integrated part of SAS Viya and is the primary user interface to import, profile, and prepare data. The Data Studio module can be accessed by clicking on the **Prepare Data** application menu at the upper left corner of SAS Viya web application window. (See Figure 4.1 below.)

Figure 4.1: Accessing Data Studio via Application Menu

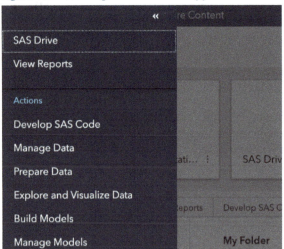

With the Data Studio window open, start a new data preparation process by selecting **New Plan**.

As a user, you have the option of using data that has already been imported via the Available tab, import new data from server-based repositories via the Data Sources tab, or import non-server based data (text files or spreadsheet) via the Import tab.

To ease ad hoc-based data discovery using data from your desktop, the Import option enables you to import a large number of data types directly from your desktop. This is a great way to import data that may have been sent to you to support ad hoc analysis quickly. The local import option supports the following types of files formats:

- Comma-separated values (CSV) files or text files

- SAS data sets (SASHDAT or SAS7BDAT)

- Microsoft Excel workbook (XLSX) files and Excel 97–2003 workbook (XLS) files

Drag and drop your data

Sometimes you get a CSV file and you just want to look at it and do some basic analysis quickly before determining whether it is of any value to you.

SAS Viya makes it extremely easy for you to import data and get going very quickly in this type of situation. You can just drag and drop a CSV file, Excel file, or SAS data set file directly onto the main design canvas of SAS Visual Analytics without even opening SAS Studio. This will automatically import the CSV or Excel file into SAS Viya, and you can start your exploration immediately. No fiddling around with buttons or options. You can bypass SAS Studio all together. You lose the ability to have more control over the import process, but you can start visualizing the data with minimal friction.

Once the import process is completed, the data set then becomes available via the Available tab. Selecting the imported data set provides you with a summary view of the imported data including the Details view, Sample Data view, and Profile view as shown in Figure 4.2.

Figure 4.2: Summary of Imported Data

The Details tab provides a high-level overview of the imported table, including the column names, total number of columns and rows, as well as the total size of the table and the last modified date.

The Sample Data tab provides a sample of the actual data imported for you to preview. Sometimes there is nothing better than "eyeballing" the data to get a feel of what the data looks like and identify any major issues. The default sample size presented is 100 rows, but you can increase that to preview more data if needed.

These summary views give you an opportunity to quickly check whether the import process was successful and whether the data were imported correctly before moving onto the data profiling stage. A little more effort and attention during this import step often means fewer problems and headaches in downstream analysis.

Data Profiling

One of the costliest mistakes an analyst or citizen data scientist can make is using bad quality data. Just because data was correctly imported does not necessarily mean that the imported data is correct and fit for purpose. The built-in data profiling capabilities of SAS Viya provide basic data profiling functionalities that enable you to quickly identify common data quality issues such as missing data, outliers, data skews, or poorly formatted data. Identifying these data quality issues early not only can prevent incorrect downstream analysis, it also gives you an opportunity to resolve some of the data quality issues during the data preparation process.

Data profile reports can be initiated and viewed via the Profile tab as shown in Figure 4.2. These reports are generated on-demand and can be saved after you initiate them. Having a historical view of the data profile reports enable you to monitor and track specific data quality metrics over time.

Run and save profile results

As well as being able to present the profile reports from previous runs in the tool itself, SAS Viya also enables you to save the profile report data into an in-memory table via the Run Profile and Save option. This option gives you direct access to the underlying profile results data. These data can then be used to develop more sophisticated data quality monitoring application or data quality dashboards.

The profile summary report provides a high-level overview of all the columns in the table. The summary statistics presented in the basic profile view includes common data quality assessment measures such as number of observations, number of missing values, mean, standard deviation, standard errors, min, and max. Depending on whether the column that you are assessing is a categorical value or numeric value, you would typically focus on and assess different data quality metrics.

Common data profile metrics to focus on include:

- For categorical or character columns: Unique, Null, Data Length, Data Type, Pattern Count, Mode
- For numeric columns: Null, Mean, Median, Min, Max, Standard Deviation

In our specific example using the World Development Indicator data set, we want to focus on a couple of columns during the profiling phase that we know we will be using in subsequent analysis. Note that some of the column names are in code (such as "SH.DYN.MORT") which we have kept to make it easier to reference back to the original data. We have included the relevant column description (such as "Mortality rate, under-5") where it makes sense to make it easier to follow the analysis.

- "Region" – Geographical Region Country (Categorical Type)
- "SH.DYN.MORT" – Mortality rate, under-5 (Numeric Type)

World Development Indicator data set
The data set that we will be using in this chapter and throughout the rest of the book is the world development indicator data set. It is a data set compiled by the World Bank, made available to the public to assist with the tracking and planning of important global initiatives such as poverty reduction. The complete data set contains over a thousand economic development indicators focused on areas such as agriculture, climate change, economic growth, education, and many other important areas. We will be using a subset of these indicators in this book to highlight the application of relevant smart data discovery techniques.

The output of the profiling process is shown in Figure 4.3 below.

Figure 4.3: Summary Data Profile Report

Column	Pattern Count	Mean	Median	Mode	Standa...	Standa...
⚠ Region	7			Europe & Central Asia		
⊕ SH_DTH_COMM_ZS		26.49	16.56		22.94	0.76
⊕ SH_DTH_INJR_ZS		9.18	8.60	7.30	4.63	0.15
⊕ SH_DTH_NCOM_ZS		64.27	70.95		23.74	0.78
⊕ SH_DYN_MORT		83.46	56.42		77.87	0.71
⊕ SH_DYN_NMRT		20.14	16.90	3.50	14.80	0.19

In terms of Region, it shows that there are seven distinct patterns in our data set, and in terms of Mortality Rate Under-5 (SH.DYN.MORT), it is showing that overall child mortality rate around the world has a mean of 20.14% and a mode of 3.5%. At a glance, it would seem that these look normal, and the metrics are what we would expect from these two columns. In order to uncover deeper data quality issues, the data profile window enables you to drill deeper into specific columns and get additional details by clicking on the specific columns name on the left-hand side.

By clicking on the Region column name, we can see more details in terms of the patterns of the variable as well as a bar chart highlighting the distribution of the categorical values as shown in Figure 4.4. In this case, we can see that most (22%) of the records belong to countries from the "Europe & Central Asia" region, which is in line with what we would expect from this data set.

Figure 4.4: Detailed Profile Report for Region

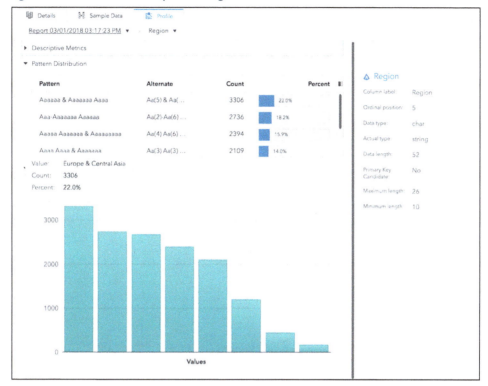

The Child Mortality Rate column (SH.DYN.MORT) is a numeric measure; therefore, we are presented with a different, detailed profile report showing a number of numeric summary statistics as well as a histogram as shown in Figure 4.5. The overall distribution as well as the median value is shown in the histogram, and once again, the shape (right skewed) of the distribution and the median value (56.42) is in line with what we would expect (and want to see) in terms of child mortality rate from around the world.

Figure 4.5: Detailed Profile Report for Child Mortality Rate Column

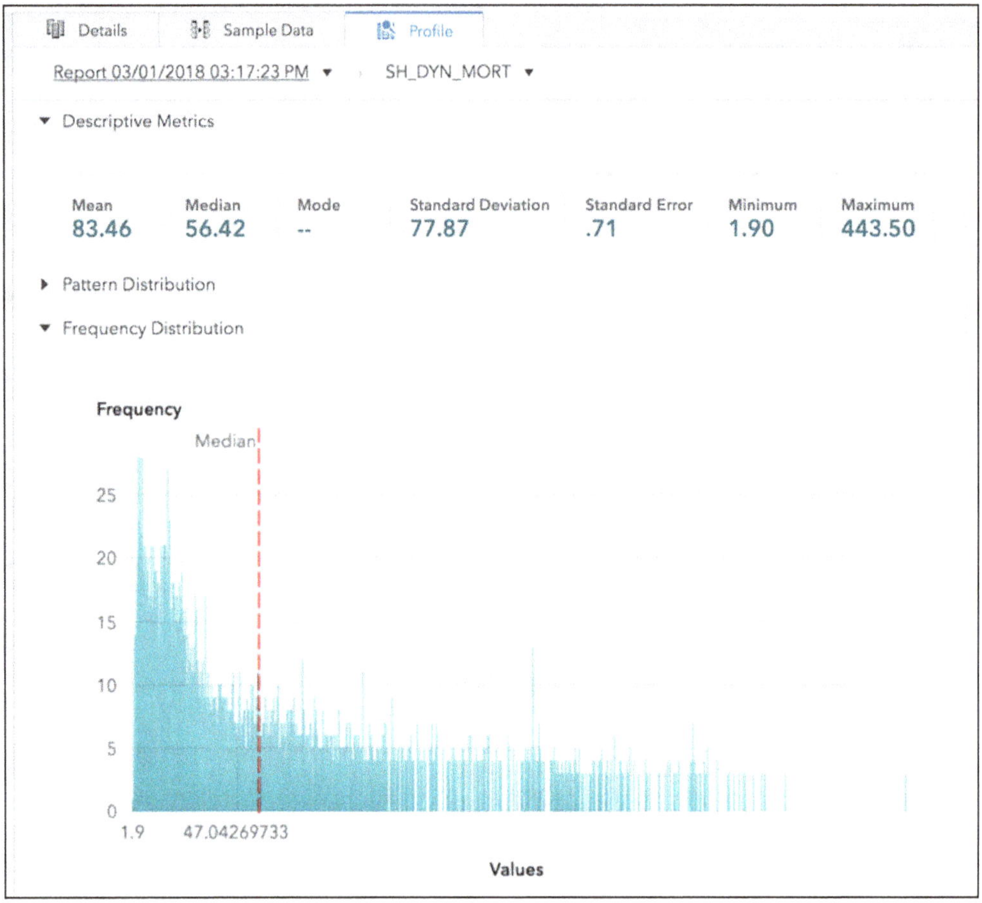

The built-in data profiling functionality of SAS Viya makes it extremely easy to validate and assess the quality of imported data before you undertake any detailed analysis. It is an important step in any data preparation process and something that you should include in your workflow especially if you are using data that you are seeing for the first time.

Data Transformation

Once data has been imported and you have assessed the quality using the data profiling functionality, you can now prepare the data by developing specific data transformation flows using a drag and drop interface.

The main Data Studio data window is split into four parts as shown in Figure 4.6. The left pane contains pre-built transformations that you can use in a data plan. The right pane enables you to view plan actions that you have defined. The top middle pane enables you to configure the selected data transformation within the current plan and the bottom middle pane enables you to view details about the select source table, including the data profile metrics that we generated in earlier steps.

Figure 4.6: Data Studio Design Environment

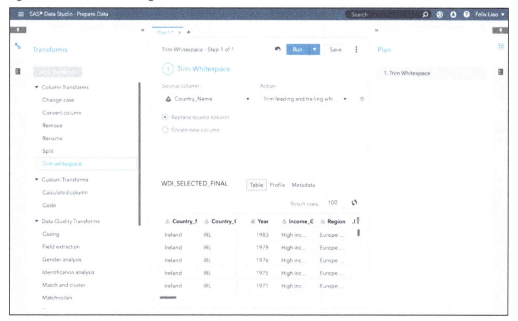

Data Plan

You can develop a data plan to perform common data transformation processes such as joining tables, appending data to a table, transposing columns, and so on. The process of building a data plan consists of developing a data plan flow using the various pre-built transformations available to you. As you add a new data transformation by double-clicking on the transformation type on the left pane, you are presented with different configuration options in the middle pane. The options available depend on the transform you have chosen. These transformations can then be chained together to form complex data transformation routines that are run together sequentially as a batch process.

SAS Data Studio provides an interactive design environment. This means that as you design your data plan and run the data plan, you are presented with a sample of the output table in the bottom pane. This interactive design paradigm enables you to quickly check your data transformation logic as you go to avoid any costly mistakes.

Figure 4.7: Comparing Source and Result Tables

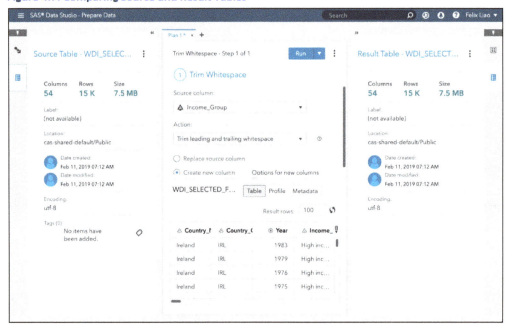

Beyond the pre-built transform steps available on the left pane, you can also develop custom transformation using code. The Code transform option enables you to build custom data transformation logics and includes support for SAS DATA step or CASL expressions.

Once you have developed an effective data plan that supports your downstream analysis, you can then save the target table as a new table or overwrite an existing table. Furthermore, you have the option of running the data plan as a one-off process or saving the data plan, which can then be scheduled to be run on a regular basis.

Get Your Data Right During Exploration

The agile nature of a data discovery process means that you often need to transform the data even further after it has been imported and loaded via a data plan using SAS Data Studio. To support these types of ad hoc, agile data preparation requirements, SAS Viya also supports basic data transformations during the data discovery phase within SAS Visual Analytics.

The goal here is to support the way an analyst wants to work and offer flexibility. The types of data preparation that you do in this context are often of an exploratory nature and experimental-focused. The preparation needs to be easy and rapid. There are, however, performance-related considerations with using either method. Generally speaking, it is more performant and efficient to use SAS Data Studio for repeatable data preparation tasks, and you should take that into consideration, especially if you are working with large data sets.

Join

SAS Visual Analytics enables you to join your data directly in the main designer environment. Coupled with the ability to easily drop and import text files within the design environment, you can quickly join and combine data sets and try new ways to visualize data and prototype models with different features.

This is done via the **New data source join…** option in the data pane, which creates a new data source that contains data from two different data sources. When creating a new data source join, Left Join is selected by default but you can also use other join types including Right Join, Inner Join, and Full Join. Furthermore, you have the option of choosing the exact join condition as well as the output columns that you want in the new target table.

Calculated Item

When it comes to using more advanced and predictive analytical techniques in a smart data discovery process, variables or features play an important role. As well as working with the set of exiting variables that have been imported from a set of source tables, it is often useful to add additional custom variables. This is a very common analytical method that is often described as features engineering. You can add new data variables within SAS Visual Analytics by clicking on **New data item** from within the data pane. We will be focusing on Calculated Item and Custom Category, which are two of the most common techniques in terms of adding new data variables.

A calculated data item is a new variable created using existing data items using an expression. For example, you could create a calculated data item called Profit, which is created by using this expression: Profit = Revenue – Cost, where Revenue and Cost are measures from an existing data source.

Another good example of a calculated item is if you want to transform an existing variable by applying a mathematical function. The logarithm function is a common transformation method used to change the distribution of a variable. It is commonly used to reduce right or left skewness of continuous variables and creates a more normal or bell-shaped distribution. A variable with a normal distribution is often more useful and required when it comes to building predictive models.

Let's try to create a new calculated item by using the logarithm function in our world development indicator data set example. Let's say we are interested in creating a new, more normally distributed variable based on the child mortality rate variable, which has a right skew as shown in the histogram in Figure 4.8.

Figure 4.8: Histogram of Child Mortality Rate Variable

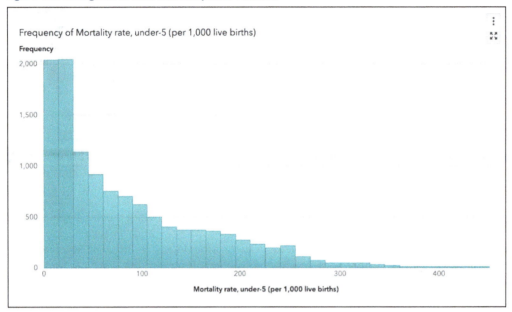

Clicking on **New data item** in the data pane and then **Calculated item** opens up the New Calculated Item window as shown in Figure 4.9. From here, you can create a new calculated item by either:

- Selecting and combining the operators and data item from the left-hand side (as shown in Figure 4.9). The GUI-driven, drag and drop option is the easiest and quickest option.

- Type out the desired expression using text (by switching from the Visual to the Text option). The Text option gives you more control and flexibility but does require you to be more comfortable with code and expressions. Both the GUI and text option can give you exactly the same results.

Figure 4.9: New Calculated Item Window

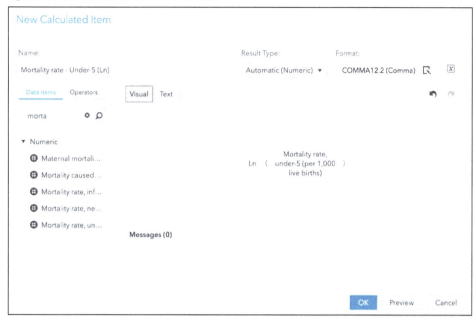

When we plot this new calculated item as shown in Figure 4.10, we have eliminated the right skewness and can now see that this new custom variable has a more normal or bell-shaped distribution, which is what we want.

Figure 4.10: Transformed Variable – Child Mortality Rate

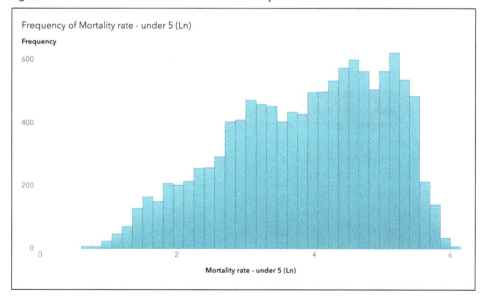

Custom Category

A custom category is a data item that you create based on either an existing category or measure data item. A custom category data item is always a category data item with an alphanumeric value. This technique is useful when you want to change or consolidate the grouping of data values in an existing data item.

Let's try to create a custom category based on the World Development Indicator data set. In this case, we will try to create a new custom category by grouping distinct values in an existing variable. Let's base our new custom category on the Income Group variable, which has four distinct values as shown in Figure 4.11.

Figure 4.11: Bar Chart of Income Group Variable

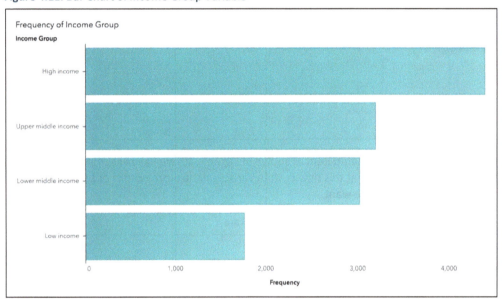

Assuming that we want a new custom category that is binary and only has two possible values, High and Low. We can easily create this new variable by clicking on **New data item** and then **Custom category** from the data pane. You will then be presented with a new custom category design window as shown in Figure 4.12 where you can name your new custom category variable and group your existing categories under new values. In our case, we will allocate the four existing income categories as either High or Low and call our new custom variable "Income Group (High - Low)" as shown in Figure 4.12.

Figure 4.12: New Custom Category Design Window

When we plot this new variable as shown in Figure 4.13, we can see that we now only have two categories instead of the original four categories. This new custom variable can then be used for situations where we want to analyze or predict a binary outcome in terms of average national income.

Figure 4.13: Income Group (High - Low) Custom Variable

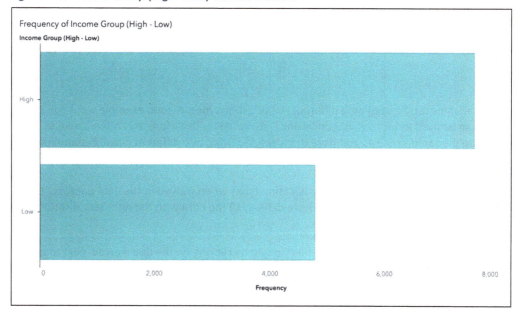

Define Variable Classification

Once you have the underlying data in the right shape, the final step before you start visualizing your data is to make sure that the variables have the right classification definitions. This is important because SAS Visual Analytics provides automation throughout the visualization process, and these automations require that you have defined the data items with the right classification type.

Furthermore, the types of data variables that you can use in specific visualizations are also controlled by the variable classification type, so having the wrong data variable classification will actually prevent you from using specific data variables in certain situations. Therefore, it is extremely important that you have defined your data with the right classification type in order to apply the correct analysis. Quite often your data might be imported as a measure type (such as a product type code) when in actual fact, it should be of categorical classification type.

SAS Visual Analytics supports a number of native variable classification types and they include:

- Category: Category types are variables that store non-numeric or non-continuous variable and are sometimes referred to as nominal variables. In our world development indicator example, country and region are examples of categorical variables.

- Measure: Measures are data items with numeric or continuous value and are sometimes referred to as interval variables. The relative size of the value is significant and used for plotting the various visualizations. Examples of measure variables in our example data set would be Life Expectancy and Child Mortality Rate.

- Date: General date format used to plot data on a time series chart. Year would be an example of a data item with Date classification type.

- Geography: Geography variables are data items that have linked graphical definitions either in the form of coordinates or boundary definitions that allow them to be plotted on a map. In our world development indicator example, country and region are examples of geographical variables.

Duplicating variables

Some variables can be used with different classification types. A good example would be Year, which can be used as both a Categorical and Date variable. Therefore, it is often useful to duplicate the same variables and have them defined using two different classification types. This can be easily done by right-clicking on a variable and then select **Duplicate**.

The variables are grouped based on classification types when viewed in the data pane, so it is often useful to first check whether you have defined all the data with the right classification type definition and re-assign them as needed.

The classification type of a variable can be changed by clicking on the down arrow next to a data item and expanding the data property window. It should be noted that when you change the classification type of a variable, it is associated with the report and not the underlying data source. Meaning that when you open the same data source in a different report, the changed classification type will not be applied. One way to reuse and share customized variable classification types is by saving and sharing data views.

Let's go back to our example data set and assume that we want to change the classification type for Year, which has been set as type Measure during the import process. We can make this change by expanding the variable property window in the data pane and then re-assign the classification type to Category as shown in Figure 4.14.

Figure 4.14: Re-assigning Classification Type for Year

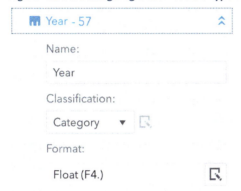

Note that the classification type options available to you in the property window depend on the chosen data item. SAS Visual Analytics tries to work out the possible classification types that you can use and offer them as options. Often, the classification type that you want might not be shown in the property window. This means that you will need to do your type re-assignment (sometimes referred to as type-casting) using the calculated item method discussed earlier.

Once we have all the variables properly defined with the correct classification type, we can then save the data view so that these definitions can be reused in other visualizations. You can open the Save Data View window by clicking on the data source action icon and then selecting **Save Data View** from the data pane. When you save a new data view, you have the option of setting the new data view as the default data view against the selected data source by ticking the **Default data view** box as shown in Figure 4.15.

Figure 4.15: Saving Data View

The process of data preparation might sound tedious and time consuming. It is however an important and critical part of a data discovery process. The more time spent and attention paid during the data preparation process, the more insights you will uncover during the actual data exploration phase.

Chapter 5: Beyond Basic Visualizations

Introduction

"If I can't picture it, I can't understand it."

– Albert Einstein

In order to truly get behind the data and uncover hidden insights and relationships, it often requires us to go beyond basic charts and visualizations such as pie charts and bar charts and take advantage of more analytical charts such as histograms and box-and-whisker plots. Going beyond basic visualizations is especially important when we want to understand relationships between many variables and uncover hidden trends that we might not even be aware of. In this chapter, we will introduce you to a number of these more analytical charts that are often misunderstood and underused. While these charts do not use predictive models or machine learning techniques (we will get that that later!), they have unique characteristics that make them powerful tools for exploration.

The charts discussed in this chapter are commonly used by data scientists before building predictive models in order to explore data and generate descriptive statistics. These charts analyze past events for patterns and provides guidance on how to potentially predict future events. They can also help guide the process of selecting inputs used in a predictive model, which is a process commonly referred to as "feature selection."

These charts are especially useful when it comes to analyzing continuous (also known as interval) variables where understanding the distribution and spread of the variable can help us understand how to approach a particular problem. Figure 5.1 shows examples of these charts, and here is a quick summary of their unique features and why you should include them in your toolkit:

- Histogram: A histogram is great for understanding the overall distribution of a continuous variable. It shows you the spread of a variable in a simple chart that looks similar to a bar chart.

- Box Plot: The box plot (sometimes called box-and-whisker plot) is a great chart to use when it comes to evaluating centrality and spread of a variable as well as identifying outliers.

- Scatter Plot and Heat Map: These are powerful charts that can highlight hidden relationships between two variables.

- Bubble Plot: The bubble plot is one of the very few charts that allows you to plot more than three variables (including the use of a time-based variable to animate the chart) and understand the relationships between them.

Figure 5.1: Examples of More Analytical Charts

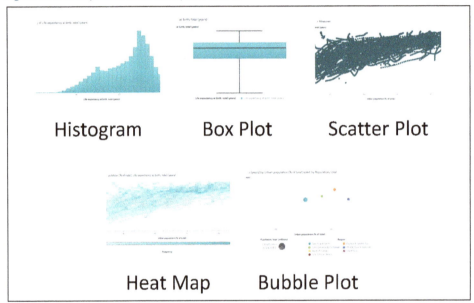

If you have never used these charts, I encourage you to venture into the unknown, learn the basics, and try them out. If you have used these charts and have been ignoring them in favor of the more basic charts, now is the time to get familiar with these powerful chart and techniques again.

Histogram

A histogram displays the distribution for a single variable and is one of the oldest charts around (first introduced by Karl Pearson in 1891!). It is considered a foundational visualization technique because the distribution of a variable helps guide how we may (or may not) use the variable for specific analytical techniques. Unfortunately, the histogram is often underused by general business users due to lack of awareness, which is a shame because it is an extremely powerful visualization technique for revealing hidden characteristics and gaining a better understanding of a numeric variable.

The histogram is drawn by "binning" up a numeric variable (binning simply means splitting up a numeric variable into equal interval groups), then counting up the number of data points in each of these groups (that is, in each bin). For example, we can bin the life expectancy of the population by counting the number of people in specific equal interval groups (1 to 10, 11 to 20, and so on).The height of the bars in a histogram indicates the count of how many data points fall into each interval group (bin). As mentioned earlier, the output of a histogram looks similar to a bar chart, but the application is quite different.

Why Distribution Matters

The histogram is extremely useful when it comes to checking whether your variable follows a normal distribution (also called Gaussian curve). Entire books have been written about characteristics of the normal distribution, which follows a classic "bell-shaped" curve, and it is the foundation of most statistical methods and approaches. A variable that follows the normal distribution is useful because the characteristics of such a variable are well-known and understood. For example, we can safely come up with important estimates for a normally distributed variable (such as 68% of values lie within one standard deviation on either side of mean) and use the variable for more advanced analysis and modeling.

Common data characteristics revealed by a histogram include:

- Distribution: Identifying the types of distribution is one the main reason for using a histogram. The common distribution types include uniform, normal/Gaussian, and Poisson.

- Skewness: Skewness is a measure of symmetry of a distribution. A distribution that is asymmetric is often described as positively skewed (in other words, pulled to the right) or negatively skewed (in other words, pulled to the left).

Let's look at our example data set using a histogram to plot life expectancy as shown in Figure 5.2.

Figure 5.2: Life Expectancy Histogram

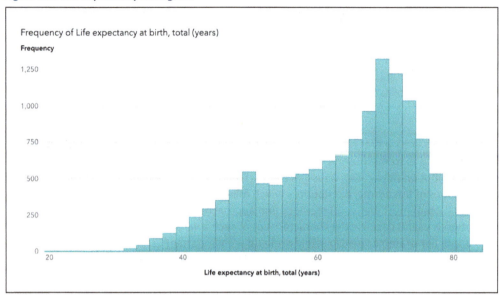

The life expectancy measure is the average lifespan a newborn can expect to have from a specific country during a particular year and is calculated using death statistics. The histogram shown in Figure 5.2 shows the distribution of life expectation for all countries from the year 1960 to 2016. From this histogram, we can note the followings about life expectancy:

- The most frequently occurring life expectancy group is 69–71 years (tallest bin).
- Overall life expectancy has a distribution that is not normal but more left-skewed with a long left tail.

The life expectancy range with the largest concentration of data points (highest bars) is what we would expect in modern, developed countries today (69–71), but the long left tail indicates that in many countries life expectancy is still low in comparison. This could be due to a large number of child deaths or infant mortality in developing countries.

SAS Visual Analytics provides a multitude of options when it comes to enabling you to modify a histogram. One of the main options is changing how bins are to be created. SAS Visual Analytics uses "system-determined values" to create bin boundaries by default, but they can be changed easily in the Options pane. The options to customize bins configuration change the granularity of the histogram, and can change how it is interpreted.

Let's assume for planning purposes that I would like to group the global life expectation values neatly into 10 groups. To do this, we set the fixed bin count to 10, and a new histogram will be generated as shown in Figure 5.3. You can find out the bin boundaries and frequency values generated by either positioning your pointer over the individual histogram bars or by maximizing the histogram and checking the tabular view below the histogram as shown in Figure 5.3.

Figure 5.3: Histogram with Customized Bins

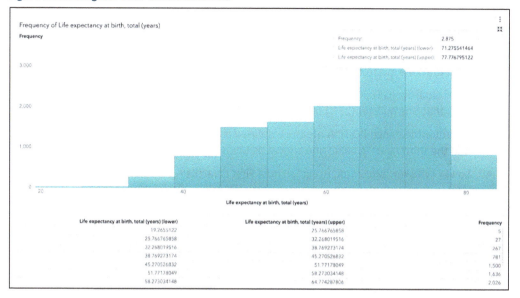

In this case, we can see that the histogram still exhibit left skewness and that the most frequent life expectancy range is now between the age of 71 and 77.

Box Plot

The box plot (or box-and-whisker box plot) is a visualization technique commonly used to understand the overall variability of your data. It extends insights that can be gained from a histogram and highlights a number of useful statistical features such as average, variance, mean, median, and quartiles. These additional statistical features provided by the box plot enable you to gain more insights in terms of skewness and central tendency. For example, analyzing the difference between the mean and median value on a box plot is another way to understand the skewness of your variable distribution.

By including additional variables, the box plot also enables you to compare distributions between different groups of data in a single chart, which can potentially highlight the relationships between variables in more interesting ways.

The box plot can display a number of statistical measures that highlight various central tendency characteristics including the following:

- Median: A key measure of central tendency. It is calculated by arranging the observations in order from the smallest to the largest values and then choosing the middle value. If there is an even number of observations, the median is the average of the two middle values.

- Mean: A measure calculated as the sum of all values divided by the number of observations. The difference between median and mean can highlight the skewness of a distribution as we will see in later examples.

- First Quartile (Q1): Represented by the bottom of the box in a box plot and is the value such that one quarter (25%) of the data lies below this first quartile value.

- Third Quartile (Q3): Represented by the top of the box in a box plot and is the value such that one quarter (25%) of the data lies above this value.

- Interquartile Range (IQR): A measure defined as the distance between the first quartile and the third quartile value and shown as the height of the box in the box plot.

- Lower Whisker: calculated using formula First Quartile – 1.5 x IQR.

- Upper Whisker: calculated using the formula Third Quartile + 1.5 x IQR.

- Min/Max: Upper and lower ends of the entire data range.

All of these statistical features are highlighted in an example box plot as shown in Figure 5.4.

Figure 5.4: Parts of a Box Plot

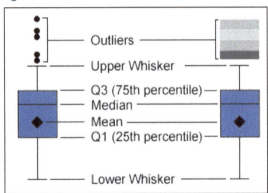

The central tendency of a variable is contrasted with its dispersion or variability and highlights how a distribution is stretched or squeezed. A common measure of dispersion is the standard deviation value, which is included and can be show on a box plot.

Let's explore the life expectancy variable using a box plot as shown in Figure 5.5.

Figure 5.5: Life Expectancy Box Plot

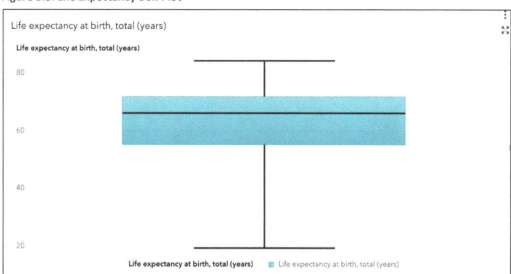

Median is displayed via the black horizontal line. As you position your pointer over the box plot, we can quickly see the following statistical features associated with the life expectancy variable:

- Median: 66.73
- First Quartile: 55.42
- Third Quartile: 72.29
- IQR: 16.87 (72.29-55.42)

From the box plot, we can make the following general observations:

- The median life span for all population is 66 years old.
- There is a relatively small interquartile range. 50% of people have a life expectancy of between 55 and 72.
- Similar to what we saw in the histogram earlier, the box plot is showing left (negative) skewness

Like all other SAS Visual Analytics charts, you can customize a box plot via the Options pane. We want to explore more using the box plot by enabling the average and selecting **Show Outliers** in the Outlier section in the Options pane. The new box plot is shown in Figure 5.6.

Figure 5.6: Box Plot with Average and Outliers

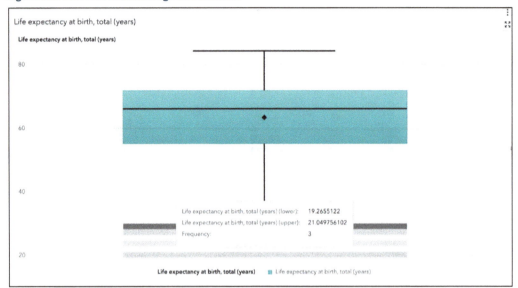

We can now see the average value displayed as a diamond icon. The gap or discrepancy between the average and median value (horizontal line) provides a rough indication in terms of the skewness of the data. This is because the calculation for the mean is affected by the magnitude of values including any extreme values (outliers). In this case, we can see the left (negative) skewness of the distribution with the average value just below the mean value on the box plot.

An outlier in SAS Visual Analytics is defined as any data point outside 1.5 times the interquartile range and is highlighted as a horizontal gray bar by default as shown in Figure 5.6.

We can now analyze the box plot with these additional statistical features:

- Mean/Average: 63.60
- Lower whisker: 30.33
- Upper whisker: 84.27
- Outlier (Lower): 18 observations, from 30.33 to 19.26

We will now add "Region" as a categorical variable into the box plot. This groups the box plot using regions and enables you to compare the characteristics of life expectancy between different regions around the world. The new box plot is shown in Figure 5.7.

Figure 5.7: Life Expectancy Box Plot Grouped by Region

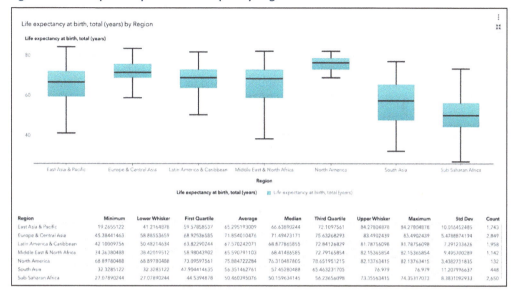

Region	Minimum	Lower Whisker	First Quartile	Average	Median	Third Quartile	Upper Whisker	Maximum	Std Dev	Count
East Asia & Pacific	19.2655122	41.2164878	59.57858537	65.295193009	66.63890244	72.1097561	84.27804878	84.27804878	10.016452485	1,743
Europe & Central Asia	45.38441463	58.88553659	68.92536585	71.854010476	71.49473171	75.63268293	83.4902439	83.4902439	5.4788874194	2,849
Latin America & Caribbean	42.10009756	50.48214634	63.82290244	67.570242071	68.877865855	72.84126829	81.78756098	81.78756098	7.291233626	1,958
Middle East & North Africa	34.36380488	38.42019512	58.98043902	65.590791103	68.41486585	72.79165854	82.15365854	82.15365854	9.495700289	1,142
North America	68.89780488	68.89780488	73.09597561	75.884722284	76.310487805	78.651951215	82.13763415	82.13763415	3.4382731835	132
South Asia	32.3285122	32.3285122	47.904414635	56.351462761	57.45280488	65.463231705	76.979	76.979	11.207996637	448
Sub Saharan Africa	27.07890244	27.07890244	44.5394878	50.460395076	50.159634145	56.23656098	73.35563415	74.35317073	8.3831092933	2,650

Note that we have also maximized the box plot (by clicking on the expand icon on the right top corner), which reveals a tabular view of the key metrics used to construct the box plot. This is useful in order to drill-in and compare the numeric values as well as the visualization itself.

With this new box plot, we can now make the following general observations:

- People from North America and Europe & Central Asia generally have the highest life expectancy.

- North America has a relatively short box plot, meaning that life expectancy is fairly consistent across the population. This is also reflected in the small standard deviation value (3.43).

- On the other hand, people living in sub-Saharan Africa have a much lower life expectancy in general and a much wider spread. The lower whisker in this region extends to the minimum across the regions, which demonstrates that this region has countries with the lowest life expectancy.

The key insight here is that region seems to be an important factor when it comes to determining and potentially predicting life expectancy. It highlights parts of the world where life expectancy is still relatively low when compared with region with more developed countries. The box plot also provides additional insights in terms of the life expectancy range and variability within each region.

Scatter Plot and Heat Map

Scatter Plot

A scatter plot is one of the best ways to visualize and inspect the direct relationship between two numeric variables. Scatter plots are also commonly used to assess and validate hidden relationships between variables.

A scatter plot plots the data points on a 2-dimensional plane (X and Y axis) and can also use color to group the points using a categorical variable. By inspecting and interpreting the patterns of data points along these two axes, you can often uncover hidden relationship between variables.

Let's explore the life expectancy variable further using a scatter plot by introducing a new variable that we believe could have a direct correlation relationship with it. We will create our scatter plot by assigning "Life Expectancy" as the Y axis and "Urban Population (% of total)" as the X axis. In this case, we want to assess any relationship between the degree of urbanization and life expectancy around the world as shown in Figure 5.8.

Figure 5.8: Scatter Plot of Life Expectancy and Urban Population (% of Total)

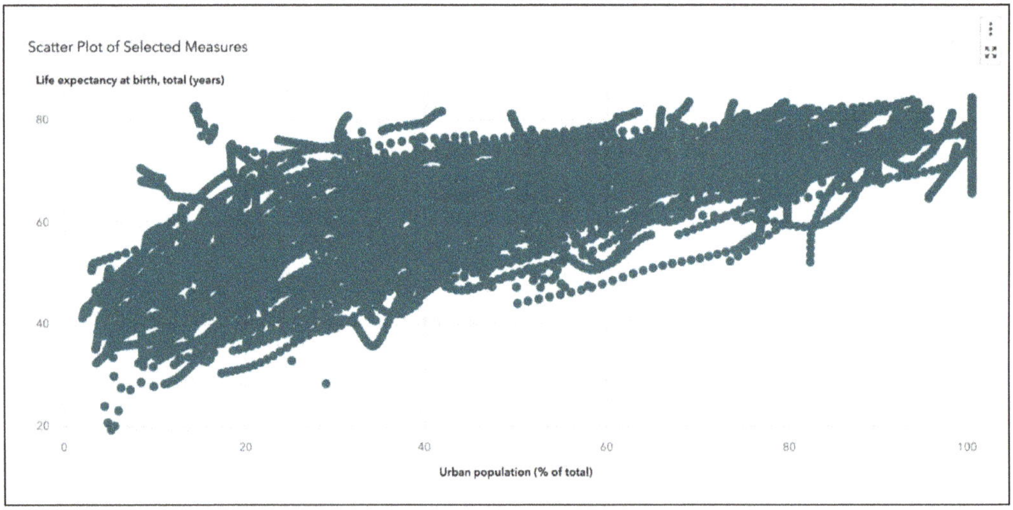

At first glance, it would seem that the degree of urbanization has a positive effect on life expectancy overall. As the degree of urbanization rises (data points moving to the right-hand side), life expectancy seems to rise (data points moving to the top of the chart). Intuitively, this would make sense as more urbanized communities tend to have better access to health care and clean water and therefore less child mortality and early death due to lack of or poor health care.

Let's assume that there is linear relationship between urbanization and life expectancy. SAS Visual Analytics allows you plot and visualize such a linear relationship by adding a fit line. A fit line gives you more insights by highlighting the slope of the linear relationship and the degree of change on the scatter plot. Various fit lines can be overlaid onto the scatter plot including linear, quadratic, cubic, and spline. The new scatter plot with the linear fit line is shown in Figure 5.9.

Figure 5.9: Scatter Plot with Linear Fit Line

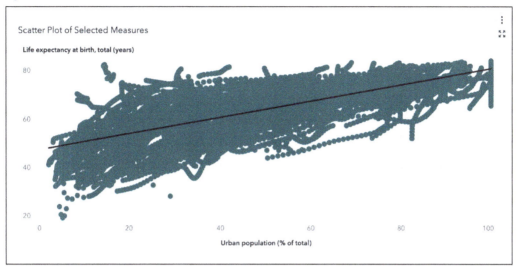

Though there seems to be a clear linear relationship, we cannot conclude that there is a direct causal effect between life expectancy and degree of urbanization. Note that the fit line and potential relationship between the variables also does not necessarily have to be linear. We will explore the topic of correlation and other fit lines in later chapters.

From the box plot earlier, we saw that region was potentially an important factor in predicting life expectancy. Let's add it into our scatter plot by assigning it to the "Color" role. The new scatter plot is shown in Figure 5.10.

Figure 5.10: Scatter Plot with Region Variable

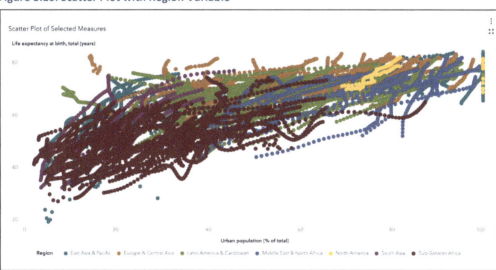

Now we can also analyze the varying life expectancy and degree of urbanization by looking at regions around the world. While the general trends in terms of the positive effect of urbanization seems to apply across the different regions, you can also see the nuances in terms of trends for individual regions. For example, we can see that people in sub-Saharan Africa (dark red) seem to have lower life expectancy (and lower degree of urbanization), which validates what we uncovered using the box plot earlier.

Scatter Plot Matrix
SAS Visual Analytics allows you to include more than three variables in a scatter plot. This is useful when you want to expand your analysis and explore potential relationships between more variables. When you use more than three variables, the chart displays a scatter plot matrix where the additional variables are added and assigned into the roles of lattice columns and rows. A scatter plot matrix then displays a series of scatter plots that highlight every possible pairing of the variables.

Heat Map

Heat maps are often used instead of scatter plots to better understand the relationships between two variables. When viewing a large number of observations in a scatter plot, it can often be difficult to interpret due the large number overlapping points. Heat maps provide a great alternative to scatter plots and allow you to identify areas with high concentration of data points by aggregating and binning the data point. Each bin covers a certain range and forms a cell on the heat map. The count of the data points in each bin is then represented by the color: the denser the number of points, the darker the color.

The other advantage of a heat map over a scatter plot is that you can include categorical variables to both axis variables (which you cannot do in a scatter plot). This means that you can use these charts to understand the relationship between more types of variables. Just to top it off, heat maps in SAS Visual Analytics also include the ability to use different frequency measures as the color indicator within each cell.

Let's consider the relationship between life expectancy and degree of urbanization using the heat map as shown in Figure 5.11.

Figure 5.11: Heat Map of Life Expectancy and Degree of Urbanization

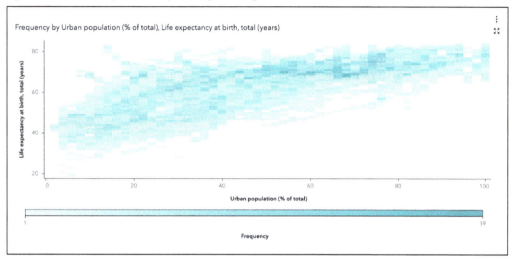

In this case, we are using the same assignments in terms of the Y-axis and X-axis variables and can see a similar pattern to what we saw earlier using the scatter plot. In addition, the heat map allows us to see high data point concentration areas more clearly because they are shown as the darker regions. In our case, we can see a cluster of points (dark region) on the right-hand side, indicating a large number of countries with life expectancies within the region of between 60 and 80 years old and degree of urbanization of between 50% and 80%.

We will now switch the color role for the heat map from frequency to population. The new heat map is shown in Figure 5.12.

Figure 5.12: Heat Map with Population Variable

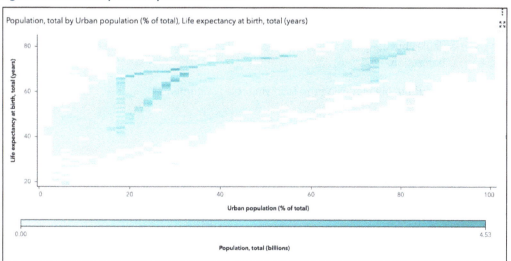

Instead of showing the density of countries in each cell, we can now see the density in each cell represented as total population. We can now see a slightly different story and can see another cluster of points (dark region) on the left-hand side of the heat map. This chart shows that there are some highly populated countries with low urban population and low life expectancy (the cluster on the left). This could be highlighting developing countries with large populations that are still lagging behind developed countries when it comes urbanization and overall population well-being.

As highlighted by the various scatter plot and heat map examples, these are great charts to use when it comes to analyzing more than one variable together. When used with the advanced configuration options, you can drill into the data sets and expose hidden relationships between variables very quickly and easily.

Bubble Plot

The bubble plot is often described as "scatter plot plus" and is a useful chart to use for analyzing multiple variables at an aggregated level. The bubble plots allow you to use a large number of variables (up to seven variables at the same time) without being too visually overwhelming. Thus, they are a great chart to use when you need to look at the big picture across multiple dimensions. When used correctly, they enable you to plot and see the interaction between many different variables in a way no other charts can.

The bubble plot enables you to visualize aggregated values along the X axis and Y axis as bubbles. The size and color of the bubbles along those two axes can be represented by additional variables. Furthermore, the bubble plot is one of the few charts that can include a date variable, which allows the bubble plot chart to be animated using the date variable and providing a very unique perspective.

Since the bubble plot is aggregating the values in each cell, it is important to first assign the correct aggregation type against the chosen variables. In our example data set, we will be using the same variables but assigning the aggregation type to average for both life expectancy and degree of urbanization.

Let's build out our bubble plot using the following role assignments:

- Y axis: Life expectancy at birth (as average)
- X axis: Urban Population (% of Total) (as average)
- Group: Region
- Size: Population (as average)

The generated bubble plot is shown in Figure 5.13.

Figure 5.13: Bubble Plot of Life Expectancy and Degree of Urbanization

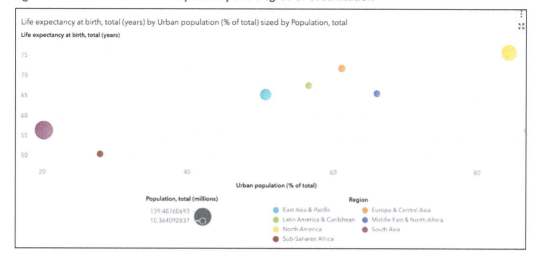

In this case, we can see that:

- On average, countries in North America and South Asia have the largest populations (largest bubble)

- Countries in North America have the highest average life expectancy and degree of urbanization (positioned in the right top corner)

- Countries in sub-Saharan Africa have the lowest average life expectancy (bubble farthest to the bottom)

This paints a picture that we are fairly familiar with and shows a world of two extremes. On one hand, we have large, developed countries in North America such as the USA. On the other hand, we have small, developing countries in sub-Saharan Africa, such as Botswana, which are less urbanized and where people have lower life expectancies.

Let's add the time dimension by assigning the "Year" variable into the role of animation in the roles section. The new bubble plot will now have an additional play bar at the bottom as shown in the diagram below. Pressing the **Play** button animates the chart and enables you to see how the bubble plot changes and evolves over the years.

Figure 5.14: Animated Bubble Plot

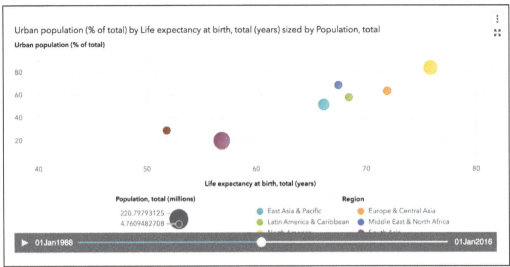

Beyond being an impressive chart, the animation allows you to see trends in terms of where the bubbles are moving (urban population versus life expectancy) and how they change in size (population) over time. In our case, the animation shows all the bubbles moving toward the right and top of the chart. This is especially the case with the bubble representing sub-Saharan Africa. This shows that as a planet, we are trending over time toward a higher population in urban areas where there are generally more employment opportunities, better health care, and higher life expectancies.

Just like scatter plots, you can add additional variables into a bubble plot and turn it into lattice bubble plot. When coupled with date-driven animation and done in a thoughtful way, bubble plots offer a compelling way to tell a powerful story by looking at trends across multiple dimensions.

Chapter 6: Understand Relationships Using Correlation Analysis

Introduction

Did you know that the amount of daily ice cream sales is related to the temperature of the day? The hotter the temperature, the more ice cream tends to be sold. This is a classic example of two phenomena that are highly correlated with each other and finding such relationships have tremendous value (such as allowing one to better plan for and sell more ice cream) and can be used in a number of business contexts.

In most data analysis, correlation refers to a linear relationship between two variables, where an increase in one variable seems to provide a consistent increase or decrease in the other variable. Correlation analysis measures and highlights this relationship and can often uncover deeper insights and help explain potential driver events.

Once a correlation has been identified, the direction and degree of correlation provide us with additional information that helps describe the nature of the relationship. Extending from the previous ice cream sales example, if the temperature goes up, the amount of ice cream sales is also likely to go up. Hence, the two variables are said to be positively correlated as shown in Figure 6.1.

Figure 6.1: Example Linear Correlation Between Ice Cream Sales and Daily Temperature

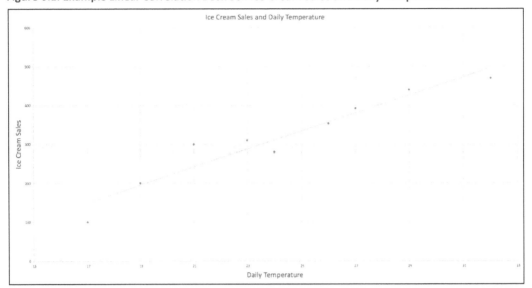

Two variables can be positively or negatively correlated. An example of negative correlation (somewhat related to eating ice cream!) would be the relationship between a person's weight and the amount of exercise he or she does. As a person exercises more, his or her weight is likely to decrease, hence exhibiting a negative correlation relationship. If we can quantify how exercise affects our weight (assuming no significant change in diet and other factors), we can act now (such as exercising more) in order to manage and control how much we weigh.

Just because two things are correlated or tend to "go together" does not necessarily mean that there is a causal relationship between the two variables. Causality often requires more in-depth analysis and knowledge of the factors involved. Having said that, correlation analysis does provide us with the first set of clues as to whether two things might have a causal relationship. The approach to identify causal relationships is a bit like what you might see on CSI, the popular TV series. Association between two variables is like finding evidence at the scene of the crime. It points us in the right direction but will typically require additional context (or motive!) and additional supporting evidence in order to lead to a fruitful conclusion. Our job when using a tool like SAS Visual Analytics is to uncover this evidence and use our domain knowledge and sound judgments in order to uncover relevant and valuable business insights.

Correlation is also commonly used as a precursor to more complex linear analysis, such as a linear regression (discussed in Chapter 9). The identification of relevant and important predictive factors to an event is often the most time consuming and complex process in a regression modeling process. Correlation analysis helps to quickly eliminate irrelevant variables. Correlation analysis is also commonly used as the first step in building a forecasting (time-dependent) model. This is especially true when we need to build a multi-variant forecasting analysis using multiple underlying variables.

As we start the process of trying to use inference and machine learning techniques to understand our world better and predict the future in the following chapters, we need to be very careful about how we extrapolate beyond the historical data analyzed. In the absence of controlled experiments, which need to be carefully designed to test for an exact hypothesis using a carefully selected set of test samples, we will always be at the mercy of the data that we have already collected. It is therefore important that we clearly understand the limitation of the data that we are using, be aware of any built-in biases, and question any assumptions used throughout the analysis. To use the ice cream sales analogy one more time, just because we see a high degree of correlation between daily temperature and ice cream sales using data that we collected from California during summer, does that mean that we can conclusively use this information to plan for ice cream sales across every state in the United States throughout the year?

By using large amounts of historical data, advanced statistical methods and powerful tools such as SAS Visual Analytics, analysts today can interpret and analyze data like never before. Smart Data Discovery, however, needs to be done carefully with a strong alignment to the business context and a deep understanding of the data. Carelessly extrapolating insights beyond the data analyzed is one potential trap that the analyst can fall into.

Correlation and Causation

It is important to use correlation analysis carefully in order to prevent harm from incorrect interpretation or conclusion. While SAS Visual Analytics provides you with a tremendous amount of power in visualizing and understanding the relationship between variables, it is still up to you to interpret the results and come up with the correct insights and conclusions. Correlation and regression analysis with SAS Visual Analytics are like one of those fancy electric power tools. They are relatively easy to use, but potentially dangerous when used improperly.

> "It ain't what you don't know that gets you into trouble. It's what you know for sure that just ain't so."
>
> –Mark Twain

One of the most common mistakes that analysts can make is confusing causality with correlation. When two variables are correlated with each other, there might well be a genuine cause-and-effect relationship (such as daily temperature and ice cream sales), but there might be other variables that are affecting both. Worse still, it might just be a coincidence!

It is up to the analyst to interpret these results and carefully draw the right conclusion through common sense and potentially testing the hypothesis using other means. Entire books have been written on the subtle ways in which statistics can be misinterpreted (or used to mislead), and we want to highlight a few of these as well here to give you some general guidelines and to avoid these pitfalls.

Extending from the ice cream sales example that we have been using, in one US city, it was found that there is a high degree of positive correlation between ice cream sales and swimming pool drownings. Should we conclude that eating ice cream causes people to drown when they go swimming? Clearly not. Upon further analysis and some critical thinking, one might conclude that this is likely a case of alternative cause or reasoning. Temperature is clearly a factor that influences both measures. One can hypothesize that as temperature goes up, people tend to buy more ice cream. It is also likely that as temperature goes up, more people go swimming, which would increase the likelihood of drowning. There will be situations where these types of incorrect conclusions seem logical, but not only are they incorrect, they are also very dangerous.

Even where a linear relationship seems to be present, we must be careful not to mix up the cause with the effect or else we might conclude, for example, that increased ice cream sales causes temperature to go up! With an absence of randomized experiments, it is important to apply critical thinking to identify true causal relationships using historical data. Here are some general guidelines that you should follow when exploring data and trying to pinpoint potential causal relationships:

1. Alternate Cause or Reasoning: If there is an alternate reason or factor (say Z) that can influence and have a direct causal relationship to both X and Y (Z causes X and Z causes Y are true), we can then reject the hypothesis of X causes Y. Failing to identify the alternative reason sometimes can lead to a completely incorrect or opposite conclusion as demonstrated by the ice cream sales and drowning example discussed above.

2. Inverse Causality: If instead of X influencing Y, we have Y influencing X, we can then reject the hypothesis of X causes Y based on inverse causality. The ice cream sales and daily temperature example discussed is a relatively straightforward example when it comes to identifying inverse causality. There will be cases where this is not so straightforward. A related point here is that we should not use input variables (sometimes referred to as explanatory variable) that might be affected by the outcome that we are trying to explain: the results will become hopelessly tangled and will make it very difficult to identify inverse casualty. At the end of the day, we should have reason to believe that our input variables affect the output variable (sometimes referred to as response variable), and not the other way around.

3. Mutual independence: Sometimes X and Y might just be correlated whilst being completely independent of each other. In such cases, we need to reject the hypothesis based on mutual independence. This is often referred to as spurious causation. A good example of this is when you try to analyze two measures that have both been rising steadily over time and likely have nothing to do with each other (for example, the number of terrorism attacks globally and the rate of economic growth in China over the last several years).

Keeping the above three guidelines in mind when you undertake correlation analysis will go a long way in helping you avoid costly mistakes and coming up with incorrect conclusions.

Now that we have established the difference between causation and correlation and the need for critical thinking, let's take a closer look at how we can do a simple correlation analysis using SAS Visual Analytics.

Correlation Matrix

SAS Visual Analytics supports the process of identifying and estimating the relationship between two or more variables through the use of a correlation matrix. In addition, there are several related SAS Visual Analytics charts and techniques that typically precede and follow the use of a correlation matrix. When combined, these techniques enable you to identify hot spots where two variables are strongly correlated, unpack and validate the nature of the relationship, and start the process of building a predictive model to quantify possible causality. We will be exploring these different techniques in more detail in the examples coming up.

The correlation matrix visualization takes several continuous variables and creates a chart that highlights where two variables exhibit a high degree of linear relationship. SAS Visual Analytics is designed to allow correlation analysis across many continuous variables simultaneously, which streamlines the process of exploring possible relationships where none are previously known.

Using the correlation matrix chart, the analyst is able to easily:

1. Identify any significant linear relationships between two interval variables
2. Identify the strength of those linear relationships
3. Identify the direction or nature of the correlations (positive or negative)

The degree to which one measure correlates with another on the correlation matrix chart is represented by the correlation coefficient. SAS Visual Analytics uses the Pearson correlation coefficient, and it is represented by the color shading of the cell within the correlation matrix chart. The darker the cell, the higher the correlation coefficient as shown in Figure 6.2. Furthermore, the exact correlation coefficient value is revealed when you position your pointer over each of the individual cells.

Figure 6.2: Correlation Coefficient Color Shading

Weak   Strong

It is useful to have a basic understanding of the Pearson correlation coefficient in order to interpret the correlation matrix effectively. Firstly, the range of the coefficient value can be anywhere from -1 to 1. A coefficient from -1 to 0 indicates a negative relationship, which means that as one of the measures increases, the other decreases, and vice versa. A coefficient of 0 shows no linear relationship. Positive coefficient values from 0 to 1 indicate a positive relationship, which means that as one measure increases or decreases, the other follows in the same direction.

Example: Explore Child Mortality Rate

Let's return to our World Development indicator data set. One of the sustainable development goals put forward by the United Nations in 2015 is around good health and well-being (Sustainable Development Goal 3). Child health and mortality rate is something that is tracked and monitored as part of this sustainable development goal. We want to track and monitor child mortality to ensure progress (progress = decreasing globally). We also want to understand factors that helps reduce child mortality rate so that we can potentially make recommendations and reduce child mortality rates across the world.

With the goal of gaining better understanding of mortality rate of children who are under the age of five, we will be analyzing various health-related measures from the World Development indicator data set (as shown in Table 6.1). We want to be able to identify characteristics of countries that have high child mortality rates through a basic correlation analysis. We will then extend this analysis using other techniques such as scatter plot and regression fit line to gain additional insights.

Table 6.1 highlights the variables that we will be using and a brief description about them.

Table 6.1: Variables Supporting Correlation Analysis of Child Mortality Rate

Variable Name	Description	Note
Mortality rate, under-5 (per 1,000 live births)	Number of child (under five years of age) deaths per 1000 live births	This is the primary (target) variable that we want to analyze. The aggregation type needs to be set to average.
Improved sanitation facilities (% of population with access)	Percentage of population with access to improved sanitation facilities	Improved sanitation facilities are defined as those designed to hygienically separate excreta from human contact. The aggregation type needs to be set to average.
Improved water source (% of population with access)	Percentage of population with access to improved drinking water sources	Improved drinking water source is defined as one that is located on premises, available when needed, and free from fecal and priority chemical contamination. The aggregation type needs to be set to average.
Physicians (per 1,000 people)	Number of doctors available per 1,000 people	The aggregation type needs to be set to average.
Hospital Beds (per 1,000 people)	Number of hospital beds available per 1,000 people	The aggregation type needs to be set to average.

Variable Name	Description	Note
Region	Major geographical regions around the world	

Note that it is important to set the correct aggregation type for the variable first before you start your analysis. The correct types are noted in Table 6.1, and Figure 6.3 shows an example where the aggregation type for **Mortality rate, under-5** has been set to **Average** using the **Options** pane.

Figure 6.3: Changing Aggregation Type of Child Mortality Rate to Average

Let Auto-Charting Guide Your Exploration

Experimentation is a big part of any smart data discovery process. To support a natural and highly responsive data exploration process, SAS Visual Analytics includes auto-charting functionalities that automatically select the relevant visualization type based on the type and combination of variables the analyst drags onto the canvas. Auto-charting speeds up the exploration process and encourages experimentation and testing of multiple variables.

When you are unsure of what charts to use, try dragging the variables that you are interested in analyzing onto the canvas and see what SAS Visual Analytics decides is the best way to visualize the data.

Leveraging the auto-charting capability, SAS Visual Analytics produces a histogram when we drag the child mortality variable onto the canvas as shown in Figure 6.4. As discussed in Chapter 5, a histogram is a great visualization to highlight important characteristics of an interval variable. When we review the histogram produced, we can see that most countries have relative low child mortality rate as shown by the two high bars on the left-hand side where child mortality rates are less than 30. The bad news is that a number of countries still suffer from relatively high child mortality rates as shown by the long tail to the right-hand side (approaching 400).

Figure 6.4: Histogram of Child Mortality Rate Variable

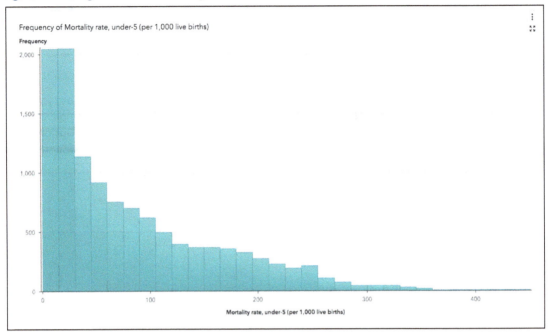

In order to get a better idea of which countries still suffer from high child mortality rates (the countries on the right-hand side of the histogram), we will now try to break down the rate of mortality by regions around the world. We will do this by changing our visualization to a box plot and add region as an additional categorical variable. The new box plot is shown in Figure 6.5.

Figure 6.5: Box Plot of Child Mortality Rate Grouped by Region

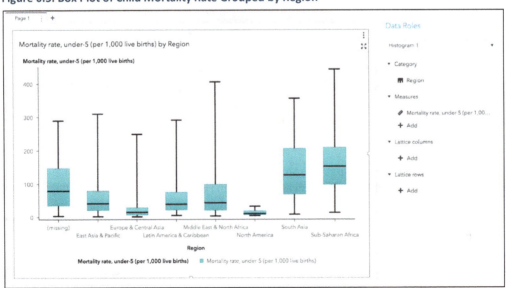

The box plot reveals several hidden details around how countries in various regions are doing in terms of child mortality rate. Countries from the sub-Saharan African region have the highest median (horizontal line) child mortality rate, closely followed by the South Asia region. Countries in the North American and European regions have the lowest child mortality and also the smallest spread in terms of child mortality rate in the interquartile range (colored box).

Unfortunately, knowing which regions have high child mortality does not directly help us in understanding the reasons for the high mortality rate. With these regions being largely consisting of developing nations, we can start to speculate about the underlying factors that contribute to the higher child mortality rate in these countries. Armed with additional health-related data, we can try to drill deeper in order to gain a better understanding of why certain countries do better in terms of having lowered child mortality and gain some actionable insights.

We will now introduce a number of new variables into the analysis that we believe should correlate with child mortality rate. These include:

- Improved sanitation facilities (% of population with access)
- Improved water source (% of population with access)
- Physicians (per 1,000 people)
- Hospital beds (per 1,000 people)

Using common sense and intuition, each of these factors should play a part in terms of the overall health and well-being of the population of a country and contribute somewhat to the health and mortality rate of children. However, we can confirm these hunches using a more scientific method. Most importantly, we would like to determine which of these factors are the most significant correlating factors when it comes to having lower child mortality rate.

Getting Your Aggregation Type Right

The ability to quickly aggregate interval values across different dimensions is one of the most powerful capabilities of SAS Visual Analytics. In addition, SAS Visual Analytics also allows the analyst to easily change the aggregation type by specifying it in the variable property in the data pane.

Getting the right aggregation type, however, requires a good understanding of the data and the relevant business context and is something that needs to be set up correctly before doing a visualization and interpreting the result.

While summation is the default aggregation type for all interval variables, summing a percentage value instead of calculating the average obviously makes no sense and needs to be avoided in order to get the correct visualization and insight.

Once we have changed the aggregation type of all the measures to average, we can then create a correlation matrix chart by dragging it onto the canvas and defining the appropriate data roles as shown in Figure 6.6.

The correlation matrix enables you to quickly visualize the degree of correlation across all the measures and identify areas of high correlation as highlighted by the darker color shading.

Figure 6.6: Correlation Matrix of Child Mortality and Related Health Measures

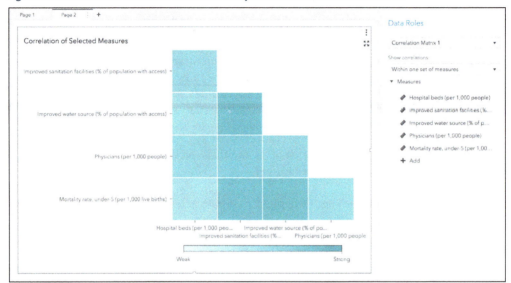

You can position your pointer over the individual cells, which will activate a pop-up window with additional details around the degree of correlation and the direction of the correlation in the selected cell. If we focus on cells with high correlation (darker colors) in our example, a number of observations can be drawn from the correlation matrix:

1. The rate of access to improved water source is highly correlated (positively) with the rate of access to improved sanitation facilities (0.8348).
2. The child mortality rate is highly correlated (negatively) with the rate of access to sanitation facilities (-0.8427).
3. The child mortality rate is highly correlated (negatively) with the rate of access to improved water source (-0.8531). This cell is highlighted in Figure 6.7.

Figure 6.7: High Degree of Correlation Between Child Mortality Rate and Rate of Access to Improved Water Source

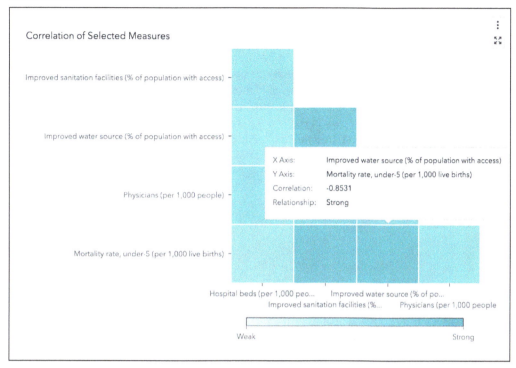

We can interpret the first observation as a reflection of the maturity of the overall drinking water and hygiene systems of a country. Countries with good access to sanitation facilities tend to also provide their citizens with good access to improved water sources. The second and third observations are potentially more valuable in explaining why some countries tend to have a high child mortality rate and others do not. Focusing in on observation three where we can see the highest degree of correlation, we can translate the observation into the following:

- Countries with high rates of access to improved drinking water sources tend to have lower child mortality rates.

- Increasing the rate of access to improved drinking water may result in lower child mortality rates.

The observation around rate of access to improved water source and child mortality makes intuitive sense and could very well lead us to some underlying causal relationships. Waterborne diseases are caused by drinking contaminated or dirty water and can cause many types of diarrheal diseases, including cholera and other serious illnesses. It is a potential reason why children in certain countries with poor access to improved drinking water sources tend to die before they pass the age of five.

This hypothesis could be further validated by looking at the causes of child death in countries with high child mortality and poor access to improved water sources. As discussed previously, while the correlation analysis has shed light on a particular area, it is now up to the analyst and the broader team to carefully work through the finding and test for other scenarios in order to get to an actionable insight.

SAS Visual Analytics provides alternative ways to visualize the correlation matrix and enables you to focus on a specific measure that you might be interested in (which, in our case, is the child mortality rate). By changing the **Show correlations** option from **Within one set of measures** to **Between two sets of measures** and defining child mortality rate as the only variable for the Y axis (as shown in Figure 6.8), we are now presented with a much more focused correlation chart focused on child mortality (as shown in Figure 6.9). This new view simplifies the chart by showing you only the correlating cells that are related to child mortality rate, making it easier to compare the correlating cells centered around the child mortality rate variable.

Figure 6.8: Simplifying the Correlation Matrix

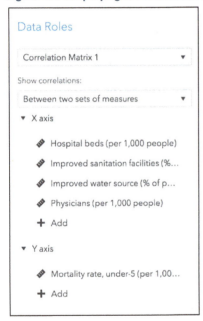

Figure 6.9: Correlation Matrix Centered Around Child Mortality Rate

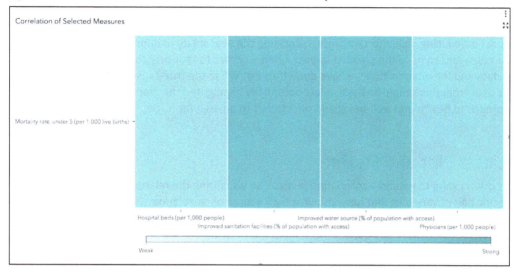

Maximizing the correlation matrix chart provides you with the additional benefit of having a tabular view of the correlation measures that display the Pearson correlation coefficient (as shown in Figure 6.10). When viewed in this manner, it leaves little doubt as to the significance of the rate of access to improved water source variable in relation to child mortality rate that has a negative Pearson correlation coefficient value of 0.8531. This indicates that as countries increase their overall population's access to improved water sources, they tend to decrease their overall child mortality rate.

Figure 6.10: Correlation Matrix Tabular View

X Axis	Y Axis	Correlation ▲
Improved water source (% of population with access)	Mortality rate, under-5 (per 1,000 live births)	-0.8531
Improved sanitation facilities (% of population with access)	Mortality rate, under-5 (per 1,000 live births)	-0.8427
Physicians (per 1,000 people)	Mortality rate, under-5 (per 1,000 live births)	-0.4925
Hospital beds (per 1,000 people)	Mortality rate, under-5 (per 1,000 live births)	-0.3937

Maximize for more details

The maximized view within SAS Visual Analytics is perhaps the most important feature available to the analyst. Not only does it allow the analyst to focus on the visualization of interest by eliminating and hiding unnecessary features, it provides additional details that allow the analyst to drill into the visualizations and ask more complex questions.

The type of detail that is made available in a maximized view varies depending on the type of visualization, but most include a tabular view of the result that can be easily exported as well as an explanation of the chart presented. Get into the habit of opening the maximized view and you will be surprised by the hidden insights that you can uncover.

At this point, the correlation matrix has revealed a number of insights about how we might potentially reduce child mortality rate in various countries. Whilst having more hospital beds, more doctors, and improved access to sanitation facilities would all seem to have a positive impact in reducing child mortality rate, increasing the availability of improved drinking water sources seems to have the greatest impact when it comes to reducing child mortality rate based on data and the analysis that we have done thus far. While the true cause of this relationship is not clear from this simple analysis, it has potentially highlighted the merits of a preventative approach to health and well-being and directed us to an area for further exploration.

Scatter Plot and Fit Line

It is often useful to extend a correlation analysis by visualizing the relationship between a specific pair of variables that have shown a high degree of correlation. This allows the analyst to ask more in-depth questions and uncover hidden insights. What is the nature of the linear relationship? Are there any outliers? And more importantly, is there a better way to model the relationship other than linear? This is where the scatter plot comes to the rescue.

The scatter plot (Chapter 5) is one of the best ways to plot and visualize two interval variables. The scatter plot plots individual data points based on two different measures, each of which is plotted against either the X axis or Y axis. The ability to see and visualize individual data points on a chart provides the analyst with a greater ability to analyze the relationship between two variables, identify potential outliers, and kick-start the process of building a linear regression model.

In our original ice cream sale and daily temperature example (Figure 6.1), we were able to fit a straight line through the data. This is commonly referred to as a linear fit line, which is a straight line that can best represent the relationship of all the data points on a scatter plot. In our specific example, the addition of the fit line on the scatter plot made it much easier for us to see the correlation pattern between ice cream sale and daily temperature.

How is the best linear fit line determined? The least squares method is generally used to calculate the best linear fit line to minimize the sum of the squares of the vertical deviations from each data point to the line (the residuals). This is not as complicated as it sounds. Each observation in our ice cream sales example has a residual (or error value), which is the vertical distance between the data point and the linear line (highlighted in red in Figure 6.11). If observations lie directly on the line, the residual value equals zero. The linear fit line is simply a straight line that results in the lowest total residual value.

Figure 6.11: Linear Fit Line with Residual or Error Bars

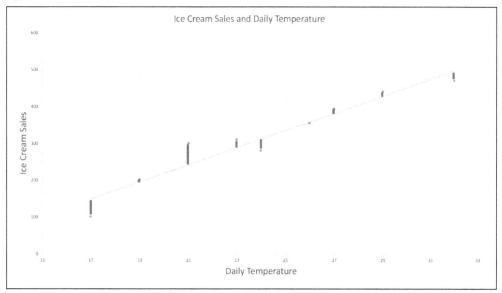

The linear fit line is actually a simple linear regression model and can be represented as an equation that takes the form of Y = a + bX. Y is commonly referred to the response variable, and X is the explanatory variable. *a* and *b* are often referred to as the parameter estimates, which are found through the regression analysis. Using our ice cream sale example for the last time, we can re-write the linear equation as:

$$\text{Ice Cream Sales} = a + b \times \text{Daily Temperature}$$

By the way, the linear fit line equation from our ice cream example turns out to be Y = -245.3 + 23.182 × X (or Ice Cream Sales = -245.3 + 23.182 × Daily Temperature)

The linear equation itself provides us with a number of useful insights through the parameter ("a" and "b") generated. The *a* parameter is called the y-intercept and has no specific meaning on its own. The *b* parameter however is known as the regression coefficient and provides us with the best estimate of the relationship between the X and Y variables. The regression coefficient is a numerical representation of the slope of the fit line generated and describes how the two variables vary together. A one-unit increase in the explanatory variable (Daily Temperature) is associated with an increase of b units in the response variable (Ice Cream Sales). Interpreting the linear equation generated from our ice cream sales fit line, it tells us that for every one-degree increase in temperature (the X variable), we should expect to sell 23.182 more ice creams (the Y variable)!

Example: Child Mortality Rate Scatter Plot

We will now focus our attention back to the example used throughout this chapter, the child mortality measure.

Let's start by first dragging a scatter plot chart onto the canvas followed by the two variables that we are interested in analyzing further: rate of access to improved water source and child mortality rate. The generated scatter plot is shown in Figure 6.12.

Figure 6.12: Scatter Plot of Child Mortality Versus Rate of Access to Improved Water Source

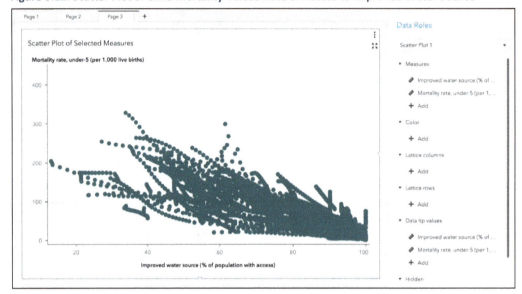

Even without adding a linear fit line, the trends of decreasing child mortality as the rate of access to improved water source increases is clear just by looking at the data points on the chart. In other words, the trend line should look like a straight line going toward to the bottom right corner.

To validate our hypothesis, we can ask SAS Visual Analytics to plot a linear fit line by going to the **Fit Line** section within the **Options** pane and selecting **Linear**. The new scatter plot with a linear fit line is shown in Figure 6.13. The linear fit line option with SAS Visual Analytics gives us the best quantified linear relationship between the two variables using the least squares method discussed earlier. It is one of many fit line options available to you including a "best fit line" option, which tries to produce a line of best fit that might involve non-linear relationships as we will explore later in this chapter.

Figure 6.13: Linear Fit Line Generated by SAS Visual Analytics

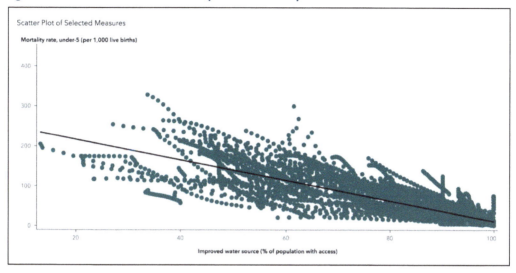

The linear fit line generated by SAS Visual Analytics provides more detail in terms of how these two measures move with one another assuming there is a linear relationship. The slope of the line and how well the points center around the fit line also enable you to assess the linear regression line in terms of goodness of fit.

In addition to the chart provided, SAS Visual Analytics also generates several useful model statistics and outputs. The maximized view provides us with this additional information as shown in Figure 6.14.

Figure 6.14: Additional Regression Line Information

Results	Improved water source (% of population with access), Mortal...
Property	**Value**
Model type	Linear
Model description	The linear fit is the straight line that best represents the relationship between two variables. If the points on the gra
R-square value	0.7278
Correlation	A correlation of -0.85 suggests there is a strong linear relationship between Improved water source (% of populati
Correlation help	A positive correlation value means that as one variable increases, the second variable increases. A negative correl
Slope	-2.5665
Function	f(x)=268.0994 - 2.5665x
Average x	83.33
Average y	54.24
Standard deviation x	17.7699
Standard deviation y	53.4587
Observations	5,987

The information table provided in the maximized view provides useful information that helps you understand the nature of the linear relationship. Some of the key properties and their descriptions are listed in Table 6.2.

Table 6.2: Key Regression Model Properties

Property	Description
Model Type	The type of model represented by the fit line. This can be Linear, Cubic, Quadratic, or PSpline.
Model Description	Characteristics of the model generated.
R-Square Value	Key statistical measure of how close the data are represented by the fitted regression line. It is also known as the coefficient of determination.
Correlation and Correlation Help	General guidelines as to how to interpret the produced correlation value.
Slope	The regression coefficient or the slope of the linear fit line.
Function	Representation of the regression line as a mathematical formula.

The R-Square value shown in the information table is commonly used as a model goodness-of-fit measure. The R-Square value represents the amount of variation that can be explained by the regression equation. It is represented as a percentage, and the larger the R-Square value, the better the model is at representing the underlying data points. Our example generated an R-Square value of 0.7278, meaning that approximately 73% of the variation in the data can be explained by the linear fit line. While the regression line certainly does not describe every observation in the data set perfectly, and it is often difficult to visually tell how well a model is in representing the underlying data set, the R-Square value is a good quantifiable metric in assessing the fit of the regression model generated.

The linear fit line that we generated supports the process of building a simple linear regression where you only need to analyze the effect of one explanatory variable (the X variable). There will be other situations where you want to analyze the effect of multiple explanatory variables in a regression analysis. The Linear Regression object in SAS Visual Analytics supports those scenarios and is covered in depth in Chapter 9.

The nature of the least square calculation used to generate the linear regression means that it is very sensitive to outliers. Outliers can affect the regression line and eventually the predicted values. One of the values of plotting the relationship using a scatter plot is to help identify these outliers. Once identified, the analyst can either drill into these outliers to identify the source of them or eliminate them all together if appropriate in order to improve the performance of the fit line. Assuming in our situation that we have identified certain data points that are outliers and can safely be removed from the analysis, this is easily done by simply selecting the data points (via region select) and then right-click to exclude the selected data points (as shown in Figure 6.15). Once removed, SAS Visual Analytics will re-calculate a new linear fit line using the remaining data points.

Figure 6.15: Excluding Outliers Using Selected Data Points

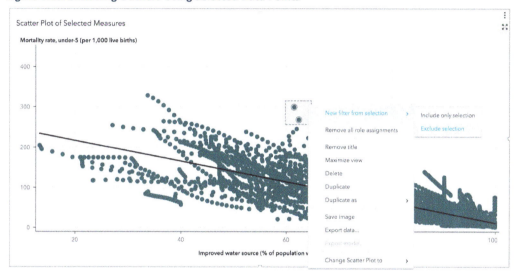

Identifying and Treating Outliers

An outlier is generally defined as an observation that deviates significantly from other observations in the data set. An outlier might be due to variability in the data itself, or it might be due to experimental or data collection error. The identification and appropriate treatment of outliers is important in order to create accurate and robust predictive models.

The involvement of SME or domain experts should play a big role in studying the outliers and ensuring that their causes are well understood. If they are indeed due to experimental or data collection errors, removing them can improve the accuracy of future analysis or models built.

While the linear fit line seems to provide us with a reasonable model in explaining the relationship between child mortality rate and rate of access to improved water sources, you can go one step further and ask SAS Visual Analytics to validate whether the linear fit line is indeed the best fit line. This is done by selecting the **Best fit** option under the **Fit line** section within the **Options** Pane. After the selection is made, all possible fit lines are calculated and the one with the highest R-Square value according to the data is selected and displayed.

In our case (as shown in Figure 6.16), the best fit line option resulted in the cubic fit line over our original linear fit line because the cubic fit line had a slightly higher R-Square value than the linear model (0.7366 versus 0.7278).

Figure 6.16: Cubic Fit Line Generated by SAS Visual Analytics

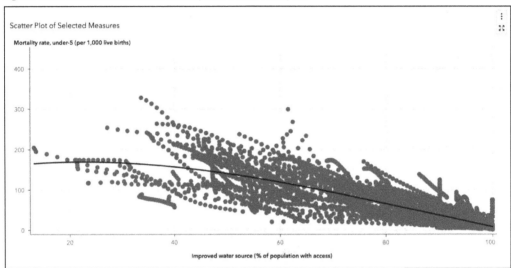

The cubic fit line gives us another way to interpret and model the relationship between the two variables. In our case, the cubic line reveals an inflection point (when the rate of access to improved water source is around 30%) where the direction of the trend line changes (from upward to downward). The details table in the maximized view (as shown in Figure 6.17) provides additional details about the cubic model as well as the cubic function itself that can potentially be leveraged to predict future child mortality rate.

Figure 6.17: Cubic Fit Line Model Information

Results	Improved water source (% of population with access), Mortal...
Property	**Value**
Model type	Cubic
Model description	The cubic fit line is used to describe the relationship between the two variables when that relationship exhibits a curve
R-square value	0.7366
Function	$f(x)=144.3862 + 2.3340x - 0.0592x^2 + 0.0002x^3$
Average x	83.33
Average y	54.24
Standard deviation x	17.7699
Standard deviation y	53.4587
Observations	5,987

To take our analysis one step further, we will now try to highlight the role that region might have when it comes to child mortality rate. To do that, we will simply add it into the existing scatter plot as a new variable with the role of **Color** as shown in Figure 6.18.

Figure 6.18: Scatter Plot with Region as Grouping Variable

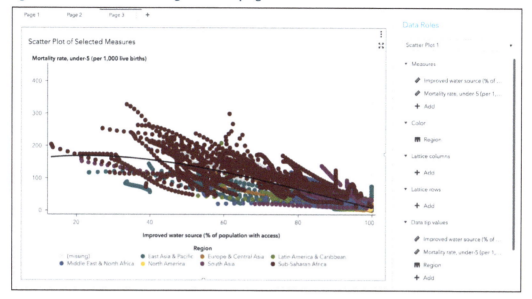

Having region as the group variable (as shown by different colors) provides us with an extra dimension in interpreting the scatter plot. A general qualitative observation of the scatter plot reveals that countries with high child mortality rate tend to be from the sub-Saharan Africa region (red data points). Countries in the Latin America and Caribbean region (green data points) have relatively lower child mortality rate but also better rate of access to improved water sources. Finally, countries from the North America region (yellow data points) tend to concentrate on the lower bottom corner where they have very low child mortality and high rate of access to improved water source. This is another data point and one that aligns well with our findings around the importance of access to improved water source.

Correlation Analysis of Categorical Variables

One of the obvious limitations of correlation is that it can only measure and analyze linear relationships for continuous variables. To measure and understand potential relationships between continuous and categorical variables, techniques like the box-and-whisker plot (Chapter 5), the scatter plot with a category overlay (as above), heat maps (Chapter 5), and crosstabulations can all be used.

Correlation analysis, when done right, opens the door to deeper understanding around potential causal relationships and enables you to start forming models that can quantify these relationships as well as predict future outcomes. It illustrates the power of an analytics-driven approach to data visualization and will form the foundation as we delve deeper into the world of visual modeling.

Chapter 7: Machine Learning and Visual Modeling

Introduction

"The signal is the truth. The noise is what distracts us from the truth."

– Nate Silver

Machine learning can be loosely defined as using a variety of supervised and unsupervised techniques to find, exploit, and optimize actions based on patterns and signals contained in the data collected. Predictive modeling is often considered the most common and useful example of machine learning.

Supervised versus Unsupervised
Machine learning models can generally be categorized as being either a supervised model or an unsupervised model. While these terms sound fancy and complicated, the distinction is really quite simple. Supervised models refer to a class of predictive models where you have a defined target that you are trying to predict such as probability of default or risk of churn. Unsupervised models, on the other hand, do not have a defined target and are very much used to find unknown patterns that might exist in your data set. Segmentation is a good example of an unsupervised model where clustering techniques are used to find patterns and natural groupings in your data set without a pre-defined target.

Statisticians and data scientists have been using machine learning algorithms and techniques to build predictive models in order to predict future events and human behaviors for a large number of years. Examples of these applications include predicting which credit card transaction is likely to be fraudulent or who is likely to default on their mortgage payments. With the vast improvement of computer power and availability of data, the applications and use cases have now expanded to all parts of the business in recent years.

In this chapter, we will be covering the foundation of building predictive models using a visual interface. Using a visual, drag-and-drop interface provided through a tool such as SAS Viya is not the only way to build a predictive model, but it can often be the fastest and most intuitive way. Visual modeling also has the advantage of making it easier for you to communicate and tell a story as you leverage predictive models to answer complex business questions. The ability to communicate and collaborate with key business stakeholders and subject matter experts when trying to build a predictive model not only removes frictions during a modeling process but also enables you to get to the root of the business issues quicker.

Predictive models can be used to generate deeper insight as well as predict future events. These models help you improve understanding of your data in ways that traditional visualizations techniques could not. The types of complex business questions that visual modeling can potentially help us answer include:

- Does seasonality matter?
- Do female customers buy larger amounts than male customers?
- Are younger customers more likely to churn?

By reviewing past events through the lens of a machine learning model, we can often learn valuable insights about past outcomes and what drove them. Extending these insights and putting them into a prediction modeling context then enables you to predict and influence future events in terms of potentially driving more sales or reducing customer churn.

Model Inference and Model Decay
It should be noted that while we can often observe trends in the data that we are analyzing, we still have the same limitations as we do with any other type of inference technique. The prediction largely depends on the historic training data set and how well it represents the broader population that we care about. Does the population that we are trying to predict change over time? If so, then the model that we built will likely suffer from model decay over time and need to be monitored and updated on an ongoing basis.

Why Visual Modeling

Building predictive models can often be a complex and time-consuming process. It typically requires a high degree of programming skills, multiple teams of people, and multiple iterations just to build and test a basic predictive model. This has limited how predictive modeling is used in organizations today. Unless you are an advanced data scientist who is competent in one of the popular data science languages (SAS, R, Python), it has been difficult to leverage these advanced, powerful machine learning techniques.

Visual modeling changes this equation and allows non-programmers and more business-aligned personnel and subject matter experts to analyze complex problem and predict future events using powerful machine learning techniques. By putting visual modeling capabilities into the hands of non-programmers, we allow subject matter expert who are deeply involved with business functions (marketing, HR, finance, customer service) to test hypotheses and ask valuable business questions.

The combination of agile, business-focused visual modeling done by the citizen data scientist community and more complex, production-class modeling done by advanced data scientists means that as a business, we can identify more valuable machine learning uses cases and iterate through more predictive model implementation cycles faster.

Approaches and Techniques

Predictive modeling is defined as the process of using data and algorithms to infer patterns that can be used to predict future events and outcomes. If the variable that we are trying to predict is a real or continuous number, the type of prediction is often described as a regression problem. If the target variable is from a limited set of discrete values or categorical values, then it is often described as a classification problem. Understanding the type of the predictive modeling first is important as it affects the types of algorithm you should use and how to assess the output of a model. We will be highlighting examples of both classification and regression problems in the following chapters.

The overall goal and approach for building a predictive model is relatively straight forward, the techniques and actual steps involved however can be quite complex. While you do not necessarily need to know the detailed mathematics behind how these algorithms work, a fundamental understanding of the various techniques and approaches is important in order to get to the right outcome and conclusion.

What Exactly Is a Prediction

A predictive model is defined as a concise representation of the input variables and output variable (target) association. A customer churn model is one such representation where we try to identify and build association between input variables such as age, gender, income, and output variable, which in this case would be the churn flag. The purpose of the training data is to generate this concise representation of the input and target variables.

The outputs of the predictive model are called predictions. Predictions represent your best guess for the target given a set of input measurements. The predictions are based on the associations learned from the training data by the predictive model as depicted in Figure 7.1.

Figure 7.1: Prediction Given a Set of Input Variables

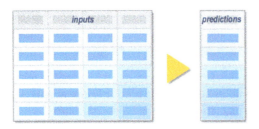

Predictions: output of the predictive model given a set of input measurements

At the heart of a visual modeling process is translating a business problem into an analytical model. The translation step involves determining which analytical methods can be applied to the objectives and problem at hand and iterating through the problem to improve the model. It is important to note that simplicity often rules. It is always advantageous to use a simpler techniques or approach if you can get a model that is good. Simpler models are easier to deploy and make the process of explaining results to key business stakeholders much easier.

Overall approach

Predictive modeling is a highly iterative process of training a model using historical data, analyzing the results, modifying the data and/or the model, and repeating. The process can be distilled down to the following key steps:

- Prepare the data and fit the model
- Tune and improve the model
- Assess/validate the model
- Compare the models and select the best

The selection of an appropriate predictive model algorithm and technique to use is often the first decision you have to make during the modeling process. Starting with a simple algorithm such as linear regression or a decision tree is often helpful in terms of understanding the data, variables, and setting a performance baseline. As you iterate through and try different techniques, the model comparison feature within SAS Viya will then help you analyze and identify the best model to use.

Techniques and Algorithms

There are a large number of machine learning algorithms available today, and this list is growing constantly. As a modeler, you need to be familiar with the common techniques in order to select the most appropriate algorithm to use. In Table 7.1, we have highlighted some of the most common techniques used by data scientists today and some of the key considerations when it comes to selecting which ones to use below.

Table 7.1: Predictive Modeling Algorithms

Technique	When to use it	Why avoid it	Note
Decision tree	Rules-based technique that is easy to interpret	Often suffers from overfitting	One of the simplest models and is also easiest to interpret
Linear regression	Well-understood and covers linear relationship well	Does not deal with missing values well	Linear regression is one of the most common regression modeling techniques and the foundation to many other advanced regression modeling techniques
Logistic regression	Extends linear regression concepts to predicting non-continuous, binary targets	Only suitable for binary classification problems	An alternative way to model for binary target variable (the other common technique being decision tree)
Generalized linear model	When the target variables have a non-continuous distribution such as Poisson or Exponential distribution	Typically requires more data transformation steps	Commonly used for credit risk modeling needed as part of loan approval processes

Technique	When to use it	Why avoid it	Note
Random forest	Generalize better than a decision tree and minimize overfitting	More compute intensive and difficult to interpret	Uses ensemble techniques to overcome the limitation of a single decision tree-based technique

The availability of these different types of approaches allow analysts to tackle different types of problems. The goal of visual modeling using these techniques is to prototype different modeling techniques quickly and then communicate and collaborate around the use of different models.

It should be noted that being able to interpret your model and deploy your model easily are two key considerations that are not obvious during the modeling process. Without compromising the accuracy of a model, a simpler model that is easier to interpret and easier to be deployed and put into production is always a better option.

Ensemble Model Rules

While we will only be covering the use of individual predictive models in this book, it is commonly agreed in the data science community that an ensemble model is often the best way to tackle complex predictive problems. An ensemble model uses and combines different models using different techniques in order to build a robust predictive model.

An ensemble approach removes the limitations of using a single algorithm and also generally does a better job of removing variance and bias in your model. There are many ensemble techniques, and they are something that you should be aware of as you try to build predictive models and work with other data scientists.

Preparing the Data for Modeling

If you ask an experienced data scientist what the most important and complex part of building a predictive model is, the answer is likely to be something related to data preparation or feature engineering. The process of selecting, modifying, and preparing your data during a model training process is fundamental to the success of your predictive model, but it is also the part that often requires the most amount of effort.

Data Partitioning

As we discussed in the previous section, the fundamental goal of building a predictive model is to develop a concise representation of input and target association. The purpose of the training data set is to generate this concise representation between the input and target variables. The model, however, needs to be general enough so that this input to output mapping extends beyond the training data set. In order to make sure you get a generalized model and avoid overfitting, you should partition your data so that you separate the data that you use to train the model and the data that you use to validate the model.

SAS Viya makes it very easy to create and assign different data partitions during the visual modeling process. You can easily create three different types of data partitions in order to train and validate your model:

- Training partition: Used to find patterns and create an initial set of models.

- Validation partition: Used to validate and select the best model from the candidate set of models.

- Test partition: Used to test and measure performance of the selected model on unseen data. The test data partition is not required, but it does provide you with additional validation options throughout the model development process.

Data partitions can be created when doing visual modeling in SAS Viya by assigning a variable to the partition variable data type. Once assigned, SAS Viya performs holdout validation, which builds the model using only the observations whose partition column value corresponds to the training data value. Next, the observations that correspond to the validation data value are then fed through the model for model validation.

You can assign the partition ID by selecting an existing variable in the data pane and then right-clicking **New partition**. You will then be asked to map the appropriate variable value to the corresponding partition type. Note that only category variables with fewer than six distinct values are eligible for use as a partition variable. See Figure 7.6 on how to assign partition variable.

Figure 7.2: Assigning Partition Variable

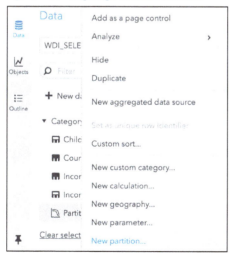

If you do not already have a partition variable candidate, the same new partition ID window can also be used to create one using random sampling techniques. Figure 7.3 is a screenshot of the data partition creation window where you can create a new partition variable on the fly.

Figure 7.3: Creating New Partition Variable

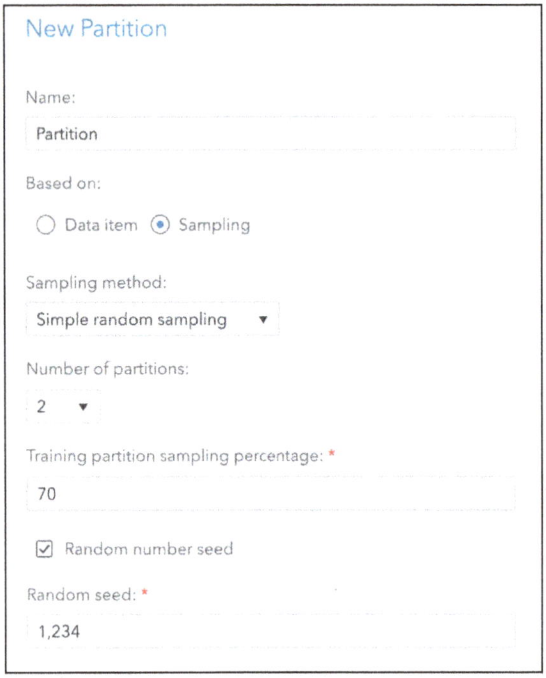

New Partition

Name:

Partition

Based on:

○ Data item ⦿ Sampling

Sampling method:

Simple random sampling   ▼

Number of partitions:

2   ▼

Training partition sampling percentage: *

70

☑ Random number seed

Random seed: *

1,234

Dealing with Missing Data

When it comes to using data to build a predictive model, the identification and treatment of missing data is critically important. This is because missing values in the training data set can reduce the fit of your model and, in some situations, can even lead to a biased model and potentially the wrong prediction. Missing values can be caused by a variety of reasons, and they often need to be carefully analyzed and dealt with during a modeling process.

Hidden Signals in Missing Data

It should be noted that missing values can also often provide relevant modeling information and can be a signal by themselves. A good example would be missing values in data collected from customer surveys. Missing values in specific fields such as age or job type in a customer survey might indicate hidden signals or intentions that we can use and should not simply be ignored. These types of situations highlight the need to understand where your data comes from and how it is collected in order to treat missing values correctly.

While some machine learning algorithms such as decision trees can inherently deal with missing data, most machine learning techniques (including most regression techniques) will suffer as a result of missing values in the predictor or input variables. By default, SAS Visual Analytics and Visual Statistics handle missing values by dropping all records that contain a missing value in any assigned role variable when you use an algorithm that cannot deal with them (think most regression models). However, most of these techniques in SAS Visual Analytics and Visual

Statistics provide an automated way to deal with missing values via the **Informative missingness** property. Selecting this property from within the Options pane tells the tool to compute and model missing values automatically. For a variable with measure types, missing values are imputed with the observed mean, and an indicator variable is created to denote missingness. For category variables, missing values are considered a distinct level. Figure 7.4 shows the informative missingness option window.

Figure 7.4: Informative Missingness Option

Informative missingness is a quick and easy way to compute missing values so that more data can be used to train the model. It provides an automated option to compute the missing values and also adds a missingness flag denoting a missing value was computed, which can also be used during the modeling process. Alternatively, you can also use custom calculation to compute and replace missing values yourself by creating a new data variable. This gives you more flexibility and control over how missing values are dealt with.

Feature Engineering

The input variables used during a model training process are often referred to as features. They have a direct impact on the quality of the model that you build, and feature engineering is where most data scientist spend a lot of their time.

Often the raw data and variables are not in a form that are suitable to train a model, but you can construct new features from the raw data that are more suitable. The process of feature engineering can also be one of the most interesting and challenging part of modeling where intuition, creativity, and "art" are as important as the technical or statistical knowledge.

Outlier Processing

Other than missing values, outliers perhaps can have the biggest impact to the quality of the model you produce. There is no universally agreed definition for outlier, but a common approach is to treat all values that lie outside two times the standard deviation value as outliers. Just like missing values, data points that are extremely large or extremely small relative to the rest of the data set can also contain important insights about your data and need to be carefully analyzed and treated appropriately.

Once you have decided that the outlier values can indeed be ignored, they are often dropped during a modeling process because they can create undesired bias in your model. In some situations, they can also be modified or flagged for inclusion during the modeling process.

Binning

Binning refers to the custom grouping of existing values in your data set. It is useful when you want to rearrange your data in order to focus on a particular aspect of your data such as a specific geographical grouping during the model process. Specific binning grouping and strategy needs to be based on the business context and modeling problem, and the actual binning can be applied to both categorical and numerical data as shown in the example in Figure 7.5.

Figure 7.5: Example of Variable Binning

```
Numerical Binning Example

Child Mortality Rate      Bin
0-200                  ->  Low
200-1000               ->  High

Categorical Binning Example

Country       Bin
Spain   ->  Europe
Italy   ->  Europe
China   ->  Asia
Japan   ->  Asia
```

A new binned variable can be easily created in SAS Visual Analytics using the custom category functionality without the need for coding.

Log Transform

Logarithmic (log) transformation is one of the most common feature engineering techniques used during modeling. It is a technique used to handle skewed data so that the distribution of your training data becomes more approximate to a normal distribution. The reason this is important is that some modeling techniques such as many regression-based modeling techniques require that your data to be close to be normally distributed. So, whenever you have a skewed distribution, you want to consider using log transformations to reduce skewness and improve the quality of the model when the underlying modeling techniques require it.

Splitting/extracting features

Often a single input variable has more than one useful feature that can be leveraged during modeling. Feature splitting or extracting enables you to extract the relevant feature from an existing variable. Date extracting is perhaps the most useful example of this. When you have date as a single input variable, it can often be useful to extract additional features out of the date variable such as day of the week or month of the year. SAS Visual Analytics is shipped with a number of transformation functions to support this requirement.

Normalize/standardize variables

It is quite common for numerical features in your training data set to have nonstandard or inconsistent ranges. A good example of this is when you have age (for example, from 18 to 80) and income variables (for example, from 10,000 to 250,000) in your training data set. Asking your machine learning model to compare them and use them to train a model can be problematic due to the different range and scale of the possible values in these two features. A difference of 10 in age can be a lot more significant compared with a difference of 10 in income in this example. Scaling techniques normalize your numeric variables so that they become identical in terms of the range (0 to 1). Scaling should not change the overall shape of the variable distribution, but it can improve the overall quality of the model as the features can be compared and analyzed using standardized metrics. Log transform and square root functions are examples of common scaling techniques that are both available in SAS Viya.

It should be noted that the effects of the outliers increase when you apply normalization/standardization techniques. Therefore, you should handle the outliers appropriately before applying normalization techniques.

Model Assessment

> "All models are wrong, but some are useful."
>
> – George E. P. Box

Once you have prepared your data and built your first model, the next obvious question is then "How good is your model?" This is where the process of model assessment comes in. Depending on the business problem and context, there is no one model assessment measure that rules them all. There are, however, a number of common model assessment measures and techniques that you should be familiar with and use in order to build a robust predictive model.

Model evaluation and assessment metrics explain the performance of a model and underpin the feedback loop used to improve your model. We will be covering some of the most common assessment metrics and techniques that are also supported by SAS Viya. These are often the most popular metrics used by data scientists and the default options used in SAS Viya. SAS Viya supports a much larger number of alternative model fit statistics that can be used during both visual modeling and code-based modeling. Please consult the product documentations if you are interested in exploring the use of these alternative metrics.

Variable Importance

Part of assessing and validating your predictive model involves understanding whether you are using the correct predictive variables in the first place. Understanding the importance of variables is not only critical when it comes to improving your predictive model, it also reveals valuable insights about the business problem that you are trying to solve. Often the value of visual modeling can come from just understanding the most important factors driving an event. SAS Visual Analytics and Visual Statistics provide a number of automated options that make this process extremely easy.

The techniques underpinning the variable importance calculation differ depending on the type of problems and algorithms used. The variable importance chart uses the information gain criterion value and is used to assess and rank the importance of predictor variables when using a tree-based technique such as a decision tree. The information gain criterion value is based on each predictor's contribution to the split in the entire tree. A larger value indicates a more significant or influential predictor variable for the model that you have created. See Figure 7.6 for an example of a variable importance chart when building a decision tree-based model.

Figure 7.6: Variable Importance Chart using Information Gain Criterion Value

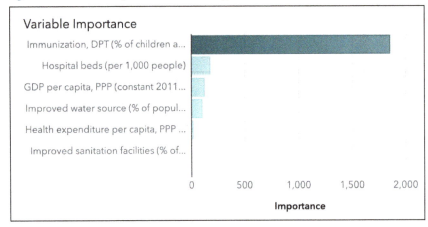

When you use a statistical-based modeling technique such as linear regression, variable importance can be assessed using the fit summary chart. The fit summary chart displays the p-value for each variable, and it is used to gauge the importance of each predictor variable. The p-value is displayed using a reverse logarithmic scale, which means that the lower the value, the longer the bar and the more important the predictor variable is. See Figure 7.7 for an example of a fit summary chart highlighting the important predictor variables in a linear regression model.

Figure 7.7: P-Value-based Variable Importance Chart

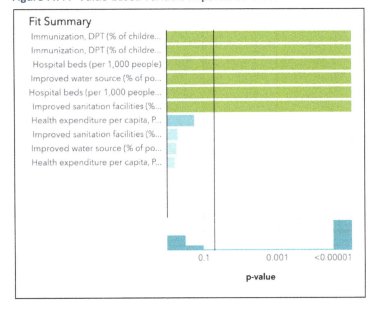

Based on the variable importance or fit summary chart, you can then eliminate variables that are not influential in order to simplify your model or add additional variables based on your understanding of the business context and situation. We will be expanding more on both of these charts using examples in later chapters.

Model Fit Statistics

When it comes to assessing the fit and performance of your model, it is important to take a holistic view and not just focus on one single measure. Depending on the specific business problem and context, you typically need to consider multiple dimensions in order to assess the fit and usefulness of your model. For example, while a model fit statistic such as R-Square is often a useful and important overall fit measure, there are situations when misclassification or false positives can have significant business impact and need to be minimized at all cost. Predictive models used in pharmaceutical or medical applications would be such an example where any incorrect prediction could have life-changing consequences (high cost of false positives).

On the other hand, when a predictive model is used in a marketing context to calculate propensity to buy a particular product, having false positives will have much lower business impact (low cost of false positives), and the assessment focus should be around overall fit and model lift.

Here are the three key model assessment dimensions that you should be familiar with and the common techniques and measures used to assess them:

- Overall fit and accuracy: The R-Square value is commonly used as the overall model fit statistic, and it is the default metric used in most regression-based models. R-Square is often described as the coefficient of determination and explains how well your selected predictive variables explain the variability in your target variable. A higher R-Square value generally indicates a better model. When a decision tree-based technique is used for classification problems, the overall accuracy of the model is commonly calculated as the proportion of data correctly classified. Alternatively, the proportion of data incorrectly classified can also be used (misclassification rate). Misclassification lets you assess the quality of your model where a low misclassification rate would indicate a more accurate and robust model.

- Sensitivity: Model sensitivity measures complements the overall model fit statistics discussed above and helps you determine whether you have produced a balanced model. Measures such as sensitivity, specificity, and the Kolmogorov-Smirnov (KS) values highlight the number of true positives and true negatives as well as how the model performs in terms of maximizing the number of true positives while at the same time minimizing the number of false negatives.

- Lift: Lift measures the performance of your model at classifying events as having the correct response. It is simply the model target response rate divided by the average response rate. A model is doing a good job if the response is much better than the average for the population as a whole (high lift value). A lift chart takes this concept further and shows you the degree of lift decline as you use the model on larger and larger data sets.

SAS Visual Analytics displays a default overall fit measure in the object toolbar on top of the design canvas. By clicking on the main model fit statistics object, you can see all available assessment measures and switch to an alternative measure as needed. See Figure 7.8 for an example of overall model fit statistic window.

Figure 7.8: Changing the Overall Fit Statistic

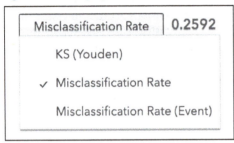

Beyond just looking at the overall model fit measure, there are also various model assessment plots available in SAS Viya that provide additional details and insights regarding the characteristics of your model and can help you assess the various model dimensions discussed above. The model assessment plot available is dependent on the types of algorithm and technique used. For classification models, the model assessment plots available include lift, ROC, and misclassification. For regression models, the default assessment plots available are the assessment plot and the residual plot.

Model Interpretability

Being able to interpret a model is an important part of model assessment. Knowing which variables are important and how they contribute to the predicted outcomes helps you communicate and explain why certain decisions are made using the model. Not only does it provide insights in terms of the predicted event, it is often a regulatory requirement in certain predictive modeling use cases such as credit risk modeling. The need to be able to interpret models helps to minimize the risk of unintended bias in the model, which can be problematic.

When it comes to model interpretability, all algorithms are definitely not created equal. While the algorithms that we are covering in this book are all well-understood and very interpretable (in terms of which variables are important and how they contribute to the prediction), certain types of algorithm such as neural networks can be more difficult to interpret and need to be used carefully in specific business contexts.

Misclassification Chart

The concept of a confusion matrix forms the foundation of many model assessment techniques for classification-based modeling. Based on the comparison between the actual outcome and the predicted outcome, a prediction is classified in one of four categories (true negative, false positive, and so on). Figure 7.9 shows an example of these four categories in a confusion matrix.

Figure 7.9: Confusion Matrix

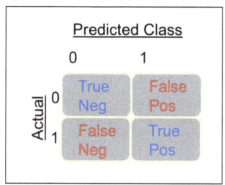

Misclassification is the default overall accuracy measure for classification when doing visual modeling in SAS Viya. It represents the ratio of incorrectly predicted observations to total observations and is essentially the sum of false positive and false negatives over the total number of observations.

The misclassification plot displays how many observations were correctly and incorrectly classified for each value and is simply a bar chart showing your predicted event ("1" in Figure 7.9 above) as one bar and the non-event ("0" in Figure 7.9) as another bar. Each bar is a stacked bar where the color indicates how many observations were predicted correctly and how many were predicted incorrectly by the model. The overall objective is to produce a model where you have low number of misclassification (the orange colored region) in all event types. Figure 7.10 below shows an example of misclassification plot for a model predicting a binary response ("High" or "NOT High").

Figure 7.10: Misclassification Plot

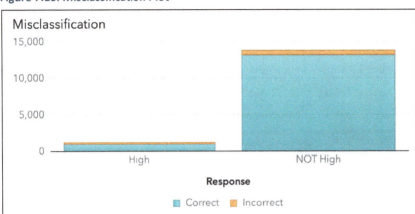

ROC Chart

Sensitivity and specificity are two important metrics related to the confusion metrics. Sensitivity refers to the number of true positives divided by the total positives cumulatively for a particular cutoff value (TP/(TP+FN)). It is a measure of how well the model correctly selects and predicts the targeted events from all the actual events.

Specificity refers to the number of true negatives divided by the total negatives cumulatively for a particular cutoff value (TN/(TN+FP)). It is a measure of how well the model correctly selects and predict non-event events from all the actual non-events.

Getting the right balance between the sensitivity value and the specificity value is an important part of the model assessment process, and this is where the receiver operating characteristic (ROC) chart comes in. The ROC chart is an assessment tool that represents the compromise between the true positive rate (using sensitivity value as the Y axis) and the false positive rate (using 1 – specificity as the X axis). The ROC chart displays the sensitivity and specificity for the entire range of cutoff values and plots how the true positive rate changes as the false positive rate changes. See Figure 7.11 for an example of an ROC chart generated in SAS Viya.

Figure 7.11: ROC Chart

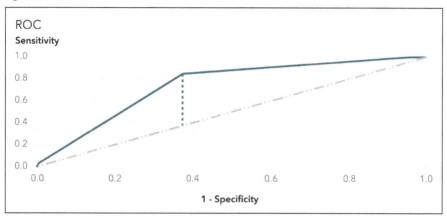

An ROC chart displays and highlights the ability of a model to avoid false positive and false negative classifications. For a perfect model, one with no false positives and no false negatives, the ROC chart would start at (0,0), continue vertically to (0,1), and then horizontally to (1,1). In most realistic modeling situations, you want to produce a good model with a very steep initial slope that levels off quickly to the right.

Lift Chart

As we mentioned earlier, lift measures the performance of your model at predicting or classifying events as having the correct response. A lift chart essentially indicates how well the model does compared to using no model (random guess). The lift is the ratio between the result predicted by the model and the result using no model.

Let's say we have the resources to target 100 potential customers in a campaign, and we want to focus on those who are more likely to respond to the campaign. There are 1,000 prospects in total, and from experience, we know 100 are likely to respond. If we select 100 at random to target, we will only likely get 10 customers who take up the offer. But what if we are able to leverage a predictive model with a lift of three? Now we can still target only 100, but this time we will likely get about 30 (three times lift) who will respond to the offer!

The lift chart takes this concept further and shows you the degree of lift decline as you use the model on larger and larger data sets. Model performance naturally declines as you use it against more data sets, and the lift chart provides a good indicator in terms of the appropriate cut off when it comes to leveraging your model to solve specific business challenges.

A lift chart plots lift on the vertical axis and depth (0 to 100%) on the horizontal axis. A lift of three indicates that your model predicted 3 times better than just selecting items at random. The default lift chart in SAS Visual Analytics displays the cumulative lift value, which is a variation on the lift measure and is calculated by using all of the data up to and including the current percentile bin. See Figure 7.12 for an example of a cumulative lift chart.

Figure 7.12: Cumulative Lift Chart

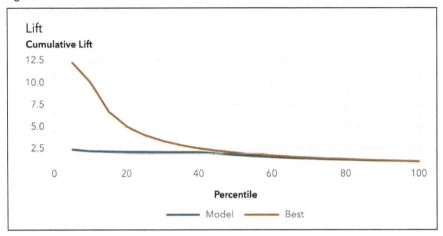

What does this graph tell you? It tells you that our model does a reasonable job until the 20th percentile where we should get at least a five times cumulative lift. The lift rate then drops gradually off toward what you would expect using random selection. The lift chart is widely used in database marketing to decide how deep in a database to go with a promotion. Also, it tells you how much response you should expect from the new target base.

Model Comparison

The model assessment object available in SAS Viya makes it very easy for you to view and compare the assessment metrics between multiple models. In order for models to be compared using the model assessment object, they must have the same data source, training and validation partition variable, response variable, event level, and Group-By variable. The model comparison object shows you a number of model assessment charts where you can compare the characteristics of the models.

It can also automatically select the best model using a comparison statistic. The default comparison fit statistic used to select the best model is the Average Square Error (ASE) rate for classification problems, but you can also change this in the Options pane. See Figure 7.13 for an example of the model comparison object.

Figure 7.13: Model Comparison Object

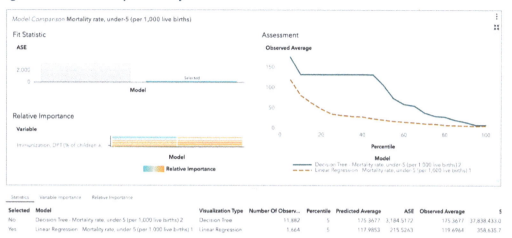

Selected	Model	Visualization Type	Number Of Observ...	Percentile	Predicted Average	ASE	Observed Average	S
No	Decision Tree - Mortality rate, under-5 (per 1,000 live births) 2	Decision Tree	11,882	5	175.3677	3,184.5172	175.3677	37,838,433.0
Yes	Linear Regression Mortality rate, under-5 (per 1,000 live births) 1	Linear Regression	1,664	5	117.9853	215.5263	119.6964	358,635.7

Improving Your Model

While just about anybody can build a predictive model using the visual modeling approach offered by SAS Viya, it is also true that not everyone can build a useful predictive model! The process of tuning and improving your model is fundamental to building a robust and useful predictive model, and it forms the iterative part of the modeling process. Not only does it require a good understanding of the business context, you will also need to bring creativity, intuition, and a good dose of curiosity. Let's go through some of the key considerations and techniques when it comes to improving your model.

Avoid Underfitting and Overfitting

The fundamental goal of machine learning is to generalize beyond the examples used in the training set. This is because no matter how much data we have, it is very unlikely that we will see those exact examples used at training time again. The most common mistake among machine learning beginners is to test the model on the training data and have the illusion of success. This means that the trained model needs to fit well with new, unseen data. Avoiding underfitting or overfitting should be the primary goal as you try to improve your model.

Underfitting your model simply means that your model is not flexible enough and is not picking up the pattern presented by the training data (blue line in Figure 7.14). On the other hand, overfitting means that your model is too flexible and picks up too much of the noise in your training data (red line in Figure 7.14) that happened only by coincidence and is unlikely to generalize beyond the training data. You want to avoid both situations and train a model that fits "just right" (green line in Figure 7.14).

Figure 7.14: How to Fit a Model

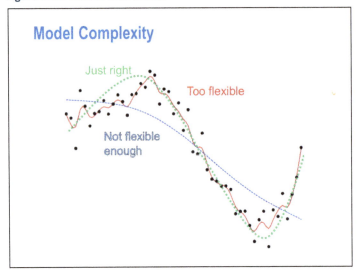

Validating and selecting a model using data that the model has not "seen" prevents the overfitting problem. This is why using partitioned data during the model training process is so important. When your initial model is 100% accurate on the training data but only 50% accurate on test data, when in fact it could have performed at 75% accuracy on both data sets, you know you have produced an overfitted model.

When you have partitioned your data into training and validation (you should!), then you can check the model fit statistics using the validation data set to avoid overfitting. SAS Viya makes it very easy to switch between the different data partitions when assessing your model. Figure 7.15 shows an example of the different model fit statistics available to you once you have partitioned your data.

Figure 7.15: Assessing Your Model Using the Validation Data Set

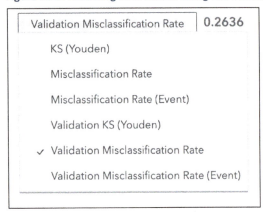

Too Good to Be True

While the goal of a modeling process is to get high level of accuracy from your predictive model, the irony is that a model that demonstrates extremely high degree of accuracy and fit against your training or variation data is often considered a bad model. An overfitted model works great against the training data set, but it often does not generalize against future data and can perform very poorly. So be cautious when you build a model that performed extremely well against your training and validation data set because often it might not perform well against future data sets.

Adding a Group-By Variable

Group-By is a technique supported by SAS Viya where you can group your data by a specific variable and create multiple models based on the grouping variable. This is useful when you know that you have distinct groups in your data set that are likely to behave quite differently. By creating multiple models using the grouping variable, you create multiple, different models that have different model parameters which better reflect the subject group that you are trying to model. A good example of this is grouping your customers by age when modeling for likelihood to churn since the factors driving churn might be quite different between a group of teenagers versus a group of seniors.

You can easily create different models using a grouping variable by adding a Group-By variable in the role option pane as shown in Figure 7.16.

Figure 7.16: Adding a Group Variable During Modeling

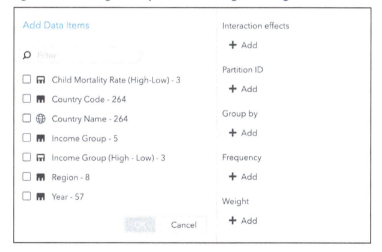

Using Better Hyperparameters

One of the reasons that you might be creating a model with poor fit is due to the use of poor hyperparameters to train the model in the first place. Hyperparameters are initialization parameters used to build a predictive model. A good example of a model hyperparameter would be the maximum number of levels allowed when building a decision tree model. Configuring and optimizing model hyperparameters can be a time-consuming process that requires good intuition developed over years of experience. Unless you are an experienced modeler, it is often a case of trial and error. Fortunately, SAS Viya offers an automated way to configure hyperparameters using machine learning techniques.

Hyperparameter autotuning is supported in a number of modeling techniques in SAS Viya visual modeling including decision trees, gradient boosting, neural networks, and others. It uses optimization techniques behind the scenes in order to automatically work out the best hyperparameters to use when building a model. As a modeler, the only thing you have to decide is how long you would like the autotuning process to run for. Figure 7.17 shows an example of hyperparameter autotuning window when you are building a decision tree model.

Figure 7.17: Autotuning Modeling Hyperparameters

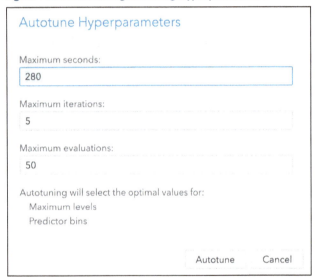

Autotuning is a great way to speed up the process of tuning your model hyperparameters and can often help you get to the best configuration and a better model much faster and easier.

Using Your Model

Generally speaking, predictive models are meant to be deployed and used in a production context. Depending on the specific environmental and business context, this might mean making a new prediction using your customers' data every month, every day, or every time they initiate a transaction. It might mean creating rules that are triggered when a customer looks at a particular item on a retail web page. Deploying a model means moving it from the data analysis and modeling environment to a production scoring environment. This can often be a complex and time-consuming process requiring the involvement of IT infrastructure specialists and data engineers.

SAS Viya offers a number of components and capabilities that allow your models to be further refined and deployed quickly and easily. These SAS Viya components include SAS Visual Data Mining and Machine Learning and SAS Model Manager, which offer model pipelines and deployment capabilities and are beyond the scope of this book.

Having said that, SAS Visual Analytics and Visual Statistics do offer additional capabilities that enable you to derive more value from your model without necessarily leveraging some of the other SAS Viya components.

First of all, once you have built a model in SAS Visual Analytics or Visual Statistics, you can easily export the model as SAS model score code via the export model function as shown in Figure 7.18. The model export and other options are available when you right-click over any of the model objects in SAS Visual Analytics or Visual Statistics.

Figure 7.18: Export Model as SAS Score Code

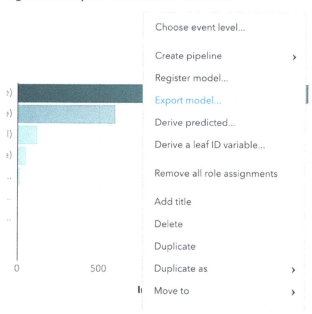

The exported SAS score code can then be executed in any environment that offers the SAS compute engine. This includes SAS Grid, SAS Studio, and SAS Enterprise Guide. Not only can other developers now execute and test your model in their own SAS environment, the exported SAS model score code can also be used to test your predictive application in a controlled or trial environment.

As well as exporting your model as model score code, you can also use the same option menu to derive the predicted value. This option enables you to leverage your model and score against the full training data set directly. The newly scored values are then added as a new column in the in-memory table for further analysis and assessment. The derived predicted value can often be used in internal reports or presentations to communicate the value of your predictive model.

We have introduced a number of fundamental concepts and techniques when it comes to doing visual modeling using SAS Viya. The best way to understand and learn these techniques further is by getting your hands dirty and tackling an actual problem, and that is exactly what we will be doing in the following chapters.

Chapter 8: Predictive Modeling Using Decision Trees

Overview

When it comes to building predictive models, it is often best to try the simplest approach first. More sophisticated techniques and algorithms are attractive, but they are usually harder to use since they typically require more configuration and fine-tuning in order to get good results. When it comes to simple machine learning models, decision trees are considered one of the simplest techniques that you should try first. Not only are they easy to visualize and build, decision tree models are also very easy to interpret and explain, which makes them great techniques to start on.

A decision tree is a classic machine learning technique that is commonly used to solve classification problems by predicting categorical targets, but it can also be used to solve regression problems predicting numeric or continuous variables, making it one of the most versatile machine learning techniques. It covers non-linear relationships, deals with missing value well, and can often be used as a variable selection mechanism for other modeling techniques.

The decision tree training process segments the data into subgroups (leaf nodes) from the original segment that contains the entire data set (root node at the top of the tree). For each subgroup, the decision tree uses a business rule to split the data at the branch. These business rules make it possible to interpret and describes the created groups (leaf nodes) in a way that is unique to machine learning models. See Figure 8.1 for an example of a decision tree predicting the color of a leaf.

Figure 8.1: A Decision Tree Model Predicting Color of Leaf

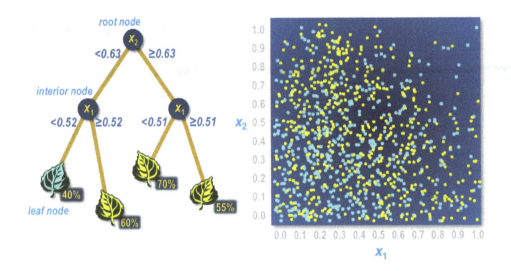

In Figure 8.1, the decision tree model on the left was created using the training data on the right, and the model can be interpreted in order to understand the importance of predictor variables (X1 and X2 coordinates in our example) as you traverse down the decision tree. The leaf nodes at the bottom of the tree include an output score (probability of leaf being teal in our example), which is used to make a prediction. Predictions are made by walking and traversing the splits of the tree until arriving at a leaf node down the bottom of the tree.

Taking the leaf node on the bottom left side of the decision tree (traversing down the branch on the left-hand side) as an example, if the X2 variable is less than 0.63 (top branch), and the X1 variable is less than 0.52 (second branch on the left), then the decision tree model is predicting that there is a 40% chance that the leaf will be teal. The sequence of the split also tells us that X2 is the more significant predictor out of the two variables given that it is the first split on top.

As you can see, decision tree models are fast to build and incredibly easy to interpret. They are also applicable for a broad range of problems and generally do not require any special preparation for your data. Let's look at how we can build our first predictive model in SAS Viya using a decision tree.

Model Building and Assessment

While a decision tree is capable of solving both classification and regression problems as we mentioned earlier, in this chapter we will highlight and demonstrate how it can be used to solve a classification problem. We will continue to use the World Development indicator data set and try to gain a better understanding of various related variables as well as try to predict the child mortality rate using a decision tree model.

Data Preparation

Child mortality is recorded as a continuous measure in the World Development indicator data set. In order to simplify our analysis of child mortality and turn this into a classification problem, we will first turn this continuous variable into a binary variable by classifying a country as having either "High" or "Low" child mortality rate for a particular year. Based on the histogram of the original child mortality rate variable that we saw in an earlier chapter (where 60 out of 1000 child mortality per year was about the overall median value), we will use 60 as the cut off value that allows us to split the data as either "High" or "Low".

See Figure 8.2 below for details of how this new custom variable (named "Child Mortality Rate (High-Low)") is created using the new custom category feature. We are essentially creating two groups ("High" and "Low") of values within this new custom category where any countries with child mortality rate greater than 60 is classified as "High" and "Low" if the rate is lower than 60.

Figure 8.2: Creating a New Custom Category

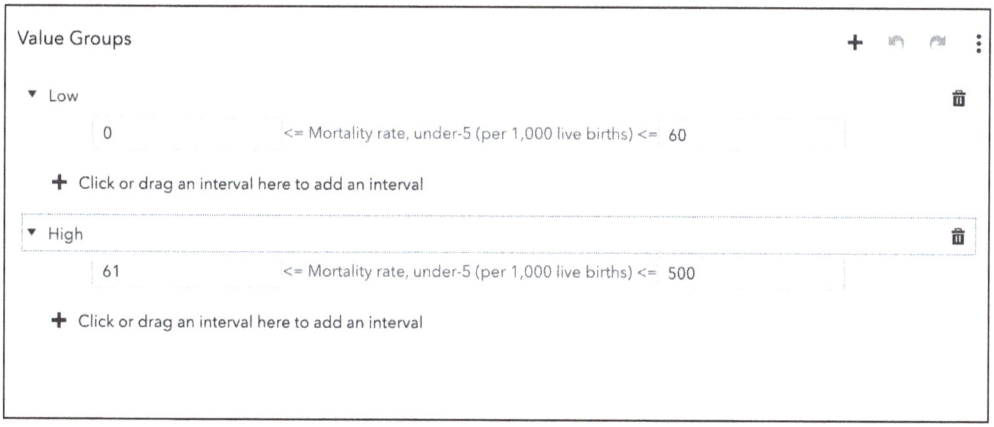

Plotting this new custom category using a bar chart (as shown in Figure 8.3) shows how the new custom variable is now fairly evenly distributed (similar number of "High and "Low") and includes some missing values that will be ignored when we try to build our predictive model in later steps.

Figure 8.3: Bar Chart of New Custom Target Variable

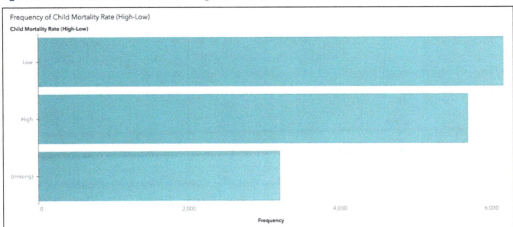

In order to avoid building an overfitted model, which can often happen when building a decision tree model, we will also need to create and apply a new data partition variable as shown in Figure 8.4. Note that we are using the random sampling technique this time and will only be creating two data partitions (training and validation split as 70/30). The use of a random seed number (1,234 in this case) allows us to re-create the same data partitions using the same seed number. In this way, we can reproduce the exact same model training and validation partitions in the future in order to validate our model training process should it be needed.

Figure 8.4: Creating Data Partitions

Predictor Variables Selection

With the data partition created, the target variable decided, and a new custom variable created, we now turn our attention to the predictor variables. When it comes to input variable selection, more is often not necessarily better. While SAS Viya includes a powerful in-memory engine that allows a large number of variables to be computed very quickly, it does not mean you should simply drag every variable into a visual modeling object. Your resultant model can be too complex if you include too many variables, particularly if you include extraneous explanatory variables with no relevance to the target variable that you are trying to predict.

As discussed in previous chapters, a number of tools in SAS Viya can help you identify suitable predictor variables to use in a predictive model. These include Box Plot and Scatter Plot as well as Correlation Matrix. These charts can be used in conjunction with the relevant business context and background in order to choose your input variables. Based on our intuition and basic understanding of global child mortality rate and potentially related health factors, we will be starting our model build with the following input variables:

- GDP per capita
- Health expenditure per capita
- Hospital beds (per 1000 people)
- Immunization, DPT (% of children ages 12–23 months)
- Improved sanitation facilities (% of population with access)
- Improved water source (% of population with access)

Building the Decision Tree Model

Let's start building our predictive decision tree model by following the steps below.

1. From the **Objects** pane, drag a **Decision Tree** object onto the canvas.
2. In the **Roles** pane, assign the newly created custom variable **Child Mortality Rate (High-Low)** as the **Response** variable.
3. Add **Child Mortality Rate (High-Low)** as a filter rule and untick **Include missing values** to remove all records with missing child mortality rate from our training and validation data set.
4. Under **Predictors**, click **Add**, and select **GDP per capita**, **Health expenditure per capita**, **Hospital beds (per 1000 people)**, **Immunization, DPT (% of children ages 12-23 months)**, **Improved sanitation facilities (% of population with access)**, and **Improved water source (% of population with access)**, then click **OK**.
5. Under **Partition ID**, click **Add**, and select the partition variable that we created earlier.

Note that the SAS Viya visual modeling environment enables you to choose the target event level when building a decision tree model. What you choose as your target event will have an effect in terms of the model fit statistics generated. In our case, we will focus on predicting countries with high child mortality rate for our analysis and choose **High** as the target event as shown in Figure 8.5.

Figure 8.5: Selecting the Target Event

Decision Tree **Child Mortality Rate (High-Low)**	(event=High)	Misclassification Rate
	Choose event level...	

Tree

You have now built your first decision tree model using default parameters, and your decision tree object should look similar to the one shown in Figure 8.6.

Figure 8.6: Decision Tree Model Predicting Child Mortality Level

The speed and ease in which you can build a decision tree model highlights the power of the SAS Viya visual modeling environment. Without a single line of code, the modeling environment has automatically split the data according to your data partition configuration, applied all the default parameters, dealt with missing values automatically, and created a decision tree structure along with all the relevant model fit statistics to help you tune and improve your model. Let's now take a closer look at what was created and what insights we can extract from our initial model.

Tree Window

The tree window is the main window displayed on the left-hand side of a decision tree object, and it is perhaps one of the most important assessment windows when building a decision tree model. The tree window contains the decision tree plot (top section) and the icicle plot (bottom section). You can also interact with the tree plot in order to tune the model interactively. See Figure 8.7 for the tree window of the model that we have created.

Figure 8.7: Tree Window from Decision Tree Model Build

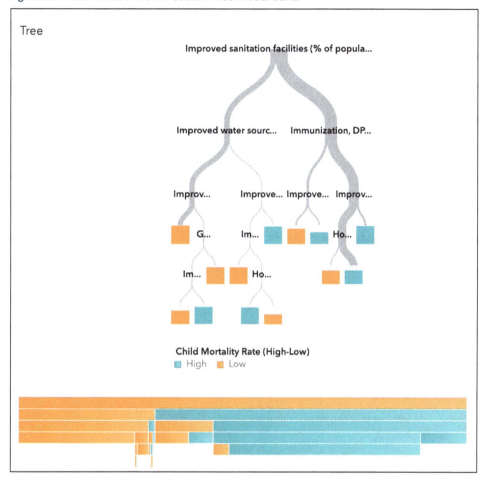

You interpret the tree plot from top to bottom to gain an understanding of the variables selected for the model as well as the splitting rules generated. If a variable is used in an earlier split near the top of the tree, then it is more important because it contributes more to the target event.

In our case, we can see that the tree plot has "Improved Sanitation Facilities", "Improved Water Sources" and "Immunization rate" near the top of the tree, meaning that they contribute the most in terms of being able to predict whether a country as having high child mortality rate. This aligns well with our prior analysis during the correlation analysis phase as well as our basic understanding and intuition around what might affect child mortality rates in developing countries.

Zoom and Using the Tree Overview Tool

For large decision tree models, the tree plot can be too big to fit on the screen. In those situations, you can use your mouse's scroll wheel to zoom in and out of the decision tree. Scroll up to zoom in, and scroll down to zoom out. In addition, the tree overview tool can be activated by clicking on the overview icon. The overview tool enables you to have more granular control around the size and location of the zoom window. See Figure 8.8 for an example of using the tree overview tool.

Figure 8.8: Activating and Using the Tree Overview Tool

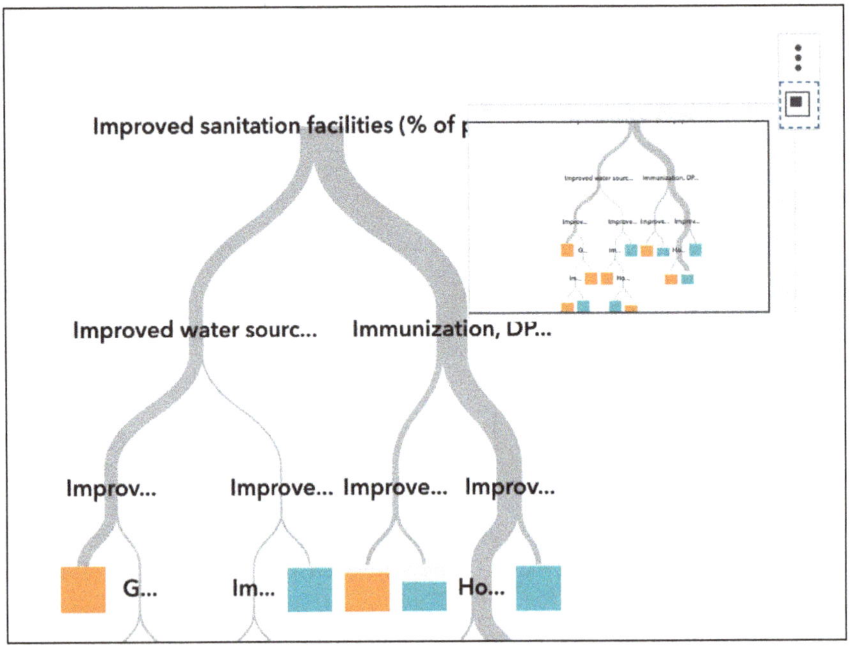

The icicle plot at the bottom of the tree window provides a simpler view of the tree structure generated and makes it easier for you to see how the data are split between the different events. The color of the nodes in both the tree plot and icicle plot indicate the predicted level for that node (Teal for "High" and Orange for "Low" in our case). When you select a node in either the decision tree or the icicle plot, the corresponding node is selected in the other location, which makes it easy for you to navigate and traverse around the tree structure.

Changing Object Layout

The default decision tree object displays all the relevant model assessment components (tree plot, icicle plot, and assessment plots) on the same page, which makes it difficult to analyze and drill into the details, especially if you have created a large tree and are working on a laptop with small screen.

Changing the plot layout from **Fit** to **Stack** in the Options pane as shown in Figure 8.9 makes it easier for you to review the objects individually by maximizing all the model fit charts. All of the objects are represented via their individual sub-pages in a stacked view that gives you more screen real estate and makes it much easier to drill into each of the model assessment charts.

Figure 8.9: Changing the Layout of the Decision Tree Model Object

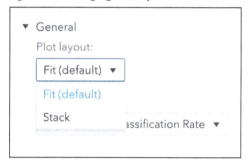

Interpreting the Model

One of the superpowers of a decision tree is that it is highly interpretable. By interpretable we mean that it is easy to use the decision tree model generated to explain the patterns uncovered from the training data set and understand how future predictions can be made.

First of all, variables near the top of the decision tree can be interpreted as being more significant and more important as we have mentioned earlier. As you traverse down the tree and the splits, you are able to build a narrative and story around the event that you are trying to predict. Figure 8.10 shows the first branch split using the "Improved sanitation facility variable" at the top of the tree. As we position our pointer over and traverse down the split, we can see that:

- The root node tells us that 48% (3,980 out of 8296 records) of all the countries have been classified as having high child mortality rate.

- "Improved sanitation facility" is the first split (from the root node) and is the most important variable when it comes to splitting the countries between ones having "High" versus not having "High" child mortality rate.

- If we traverse and position our pointer over the node on the right-hand side of the tree, we can see that this is a subgroup of countries that have "Improved sanitation facility" values that are less than 61.04 or missing. It is also a group that tends to have a high child mortality rate with 67% of records (3852 out of 5735 records). When compared with the overall child mortality rate (48% at the root node), we can interpret this as when countries have relatively low access to sanitation facilities, they tend to have higher child mortality rate.

- Alternatively, if we traverse and position our pointer over the node on the left-hand side of the tree, we essentially get the opposite side of the story. We can see that this is a group of countries that have "Improved sanitation facility" values that are greater than 61.04 or missing. It is also a group that is less likely to have a high child mortality rate with 2.41% of records (55 out of 2282 records). We would interpret this as when countries have relatively high access to sanitation facilities, they tend to have a low child mortality rate.

- As we traverse down both the left and right nodes, we can also see what the next most important variables are. "Immunization rate" is the next most important factor as we traverse down the right-hand side of the tree, and "Access to improved water source" is the next most important factor as we traverse down the left side of the tree.

Figure 8.10: Interpreting the First Split of the Decision Tree

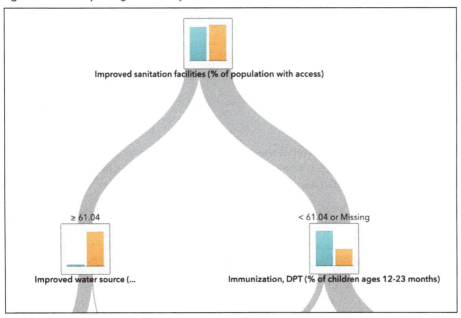

The narrative provided by the decision tree aligns well with some of the earlier analysis that we have done and our intuitions. As we traverse down the tree, review the overall structure of the tree and analyze the split and the variables selected by the tree. You can tell a powerful story around the target variable and how the input variables correlate to the target variable (high child mortality in our case).

When you eventually traverse down to the leaf nodes down at the bottom of the tree, you end up with a set of business rules that enable you to potentially predict the child mortality outcome for a particular country. Looking at leaf node 12 shown in Figure 8.11, we can see that this leaf node includes countries that have low rates of immunization (<84.3 or missing) and low rates of access to improved sanitation facilities (<46.43). This is also a leaf node that includes mostly records that have high mortality rate (95.39%). As we apply new data to this model and score them using our newly built decision tree model, any data that has the profile of input variables that satisfy these conditions will fall into this leaf node and be classified as having high child mortality rate based on the decision tree model.

Figure 8.11: Interpreting the Leaf Node

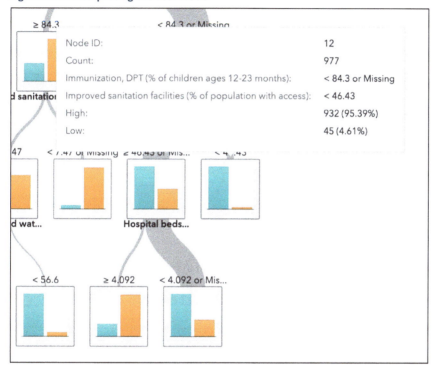

Overall Fit and Assessment Plots

Our first decision tree model has given us a lot of insight in terms of what might be the important and relevant factors when it comes to child mortality rate, but if we want to use this model for prediction purposes, we need to make sure that we have a robust and accurate model that generalizes well beyond the training data. This is where overall fit statistics and assessment plots come in.

Let's start by looking at the overall model fit statistics. In our specific example, we will focus on the misclassification rate, which gives us a reasonable indication of how accurate the model is from the perspective of minimizing incorrect predictions. Without tweaking any of the model parameters and just using the default modeling parameters, we can see that we have produced an initial model with a misclassification rate of 0.179, meaning that for every 100 records predicted, it made 17 incorrect predictions when compared with the actual outcomes. Note that this assessment statistic is generated using the training data set, which is normally a bad practice because it can hide a model that is overfit. The SAS Viya visual modeling environment enables you to switch the overall fit statistic to use the validation data set instead as shown in Figure 8.12. In this case, the misclassification rate using the validation data set turns out to be 0.1713, which is close to the misclassification rate generated using the training data set and what you would like to see in most modeling situations because it tells you your model generalized well beyond the training data set.

Figure 8.12: Switching Overall Model Fit Statistics

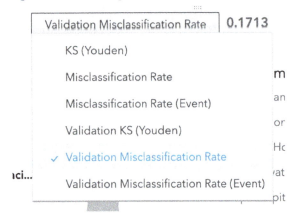

Misclassification Plot

Speaking of misclassification, the misclassification plot shown in Figure 8.13 provides more details around the predictions made and can help you assess whether you have produced a balanced model in terms of minimizing false positives and false negatives.

Figure 8.13: Misclassification Plot

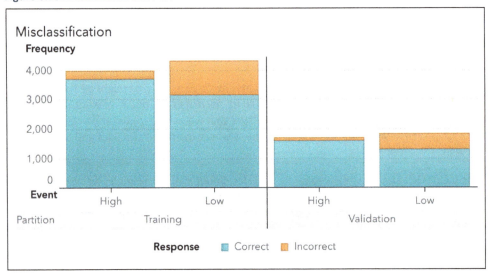

When you have partitioned and use partition data in your model, the misclassification plot produced contains both the misclassification plot generated using the training data set as well as the one generated using the validation data set as shown in Figure 8.13. In our case, we can see that the training data set and validation set exhibit similar characters in terms of having a relatively large number of false negatives (countries with high child mortality that are incorrectly classified as Low). This is indicated by the large orange regions in the misclassification chart in our example. Having high rates of false negatives means that we can potentially miss countries that will have high rate of child mortality (by incorrectly predicting them to have low mortality). This is not an ideal scenario in our use case and is probably something that we will need to focus on addressing as we iterate and try to tune and improve the model going forward.

Variable Importance Chart

The variable importance chart generated and shown in Figure 8.14 re-enforces what we have already learned via the decision tree structure itself in terms of the relative importance of the variables used. It highlights the "Improved sanitation facilities" variable as being clearly the most important factor within our decision tree model. It also highlights "GDP per capita" and "Health expenditure per capita" as having very low importance in this model, and these variables are something that we should consider removing if we want to simplify our model.

Figure 8.14: Variable Importance Chart

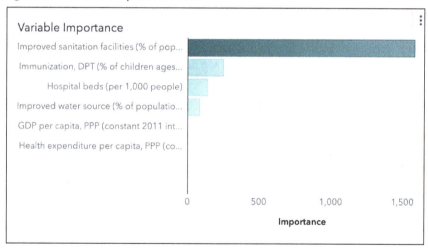

Lift Chart

The model line in the cumulative lift chart provides an indication of how much better the model is able to predict the desired outcome (countries with high child mortality) versus random selections. It is an important chart to use in determining the effectiveness of classification models, and it is often an area where data scientists focus in order to get incremental improvements during the model tuning process.

As shown in Figure 8.15, a lift rate of 1.56 in our validation data set means that our model is able to predict 1.56 times better than just selecting countries at random when it comes to choosing countries that will have high child mortality rates. The model is also able to sustain this lift rate until the 60th percentile using the validation data set but then the lift rate drops off gradually, which is what normally happens as we try to predict using larger data sets. As we continue to tune and improve our model, what we would like to see is a lift chart showing a higher lift rate and one that sustains a superior lift rate for larger percentage of the data.

Figure 8.15: Cumulative Lift Chart

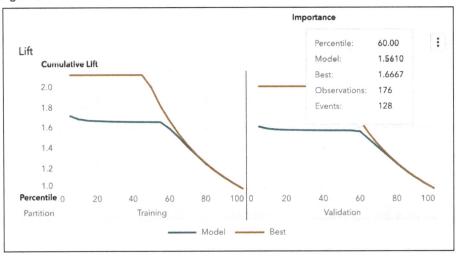

The initial build of the model is often an important step in establishing the model performance baseline benchmark using default parameters. It guides and points us in areas that need further refinements and potential improvements, which we will take a look at in the next section.

Tuning and Improving the Model

With the goal of being able to reliably predict the rate of child mortality in future years, the next step of the modeling life cycle involves tuning and improving our model through an iterative process of incremental adjustments and improvements.

The initial set of fit statistics and assessment plots have established a set of baselines and provided us with some useful guidelines in terms of where to focus for incremental improvements. In our case, we want to focus our efforts by looking at ways to:

- Improve the model performance by reducing the overall misclassification rate
- Increase the cumulative lift rate of the model
- Reduce the overall complexity of the model

With the help of the relevant business domain knowledge and some intuition, we will now look at some of the steps that we can take in order to improve the robustness and performance of our model.

Adding and Removing Predictor Variables

Adding relevant predictor variables is perhaps the easiest and most obvious way to improve our model. Doing this thoughtfully without adding to the complexity of the model is a great first step when it comes to tuning and improving your model. We will start by adding two additional predictor variables that should relate to the overall population health and child mortality:

- Physician (per 1,000 people)
- Urban population (% of total)

Based on the variable importance chart that we saw earlier, we also saw that we had variables that do not seem to contribute to the model at all and we would also like to remove at this stage so that we can produce a simpler model. Let's update our decision tree model by also removing the following variables:

- GDP per capita
- Health expenditure per capita

A new model along with a new variable importance chart is created, which is shown in Figure 8.16. With the change of predictor variables, we can also now see an improvement in our model performance as the overall misclassification rate has dropped from 0.1713 to 0.1209.

Figure 8.16: New Variable Importance Chart

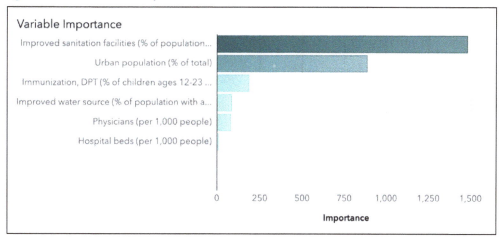

The new variable importance chart tells us that the newly added degree of urbanization ("Urban population % of total") variable turns out to be a strong predictor (second on the chart). The physician variable ("Physicians per 1000 people") is also an important predictor but to a lesser degree. But more importantly, we are moving in the right direction in terms of decreasing the misclassification rate and improving the overall performance of our model.

In one single step, we have identified new predictor variables, improved the model accuracy, and at the same time simplified the model by removing non-contributing variables. Variable selection is a powerful way to improve and tune your model, but it is something that often needs time, intuition, and a deep understanding of the business context. It is also something that needs to be done in a thoughtful way as not to overcomplicate the model or create a model that does not generalize well beyond the training data set. We have really only scratched the surface in covering this important area of modeling, and it is something that you should build on as you advance your modeling journey.

Dealing with Missing Values

As we mentioned earlier, the decision tree algorithm is a fairly robust algorithm that generally deals well with missing values in the training data set. Having said that, there are ways to tune and modify your model in terms of dealing with missing values that could potentially improve your model.

The default missing value assignment strategy within the SAS Viya visual modeling environment is **Use in search**, which means that any missing values in the training data set are treated as separate values. This might or might not be the best strategy based on the types of problem you are trying to predict and the characteristics of your training data set.

One of the other common missing value assignment technique supported is **None**, which ignores any observations that has a missing value. We know that we have a fairly sparse training data set with lots of missing values, so this is not an ideal strategy because selecting it will result in a large number of records being dropped. In our case, we will try switching the missing assignment strategy to **Popular** as shown in Figure 8.17. The Popular strategy assigns records with missing values to the child node with the most observation during a split and can be an effective missing assignment strategy in certain situations. This strategy can be effective if you believe the observations with missing values in your data set behave largely like the rest of the data set.

Figure 8.17: Missing Value Assignment Strategy

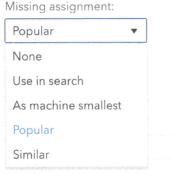

In our case, changing the missing value assignment strategy to **Popular** actually resulted in a slightly worse model with the misclassification rate increasing from 0.1209 to 0.1287. As a result, we will revert to our original missing strategy of **Use in search**. This step highlights the trial and error nature of the model tuning process. As you try different parameters and configurations, you do not always get the desired outcome and improve your model, which is perfectly normal. It is very much about experimenting with different approaches, identifying where improvements come from, and then focusing your effort on those areas for incremental gains.

Tuning the Hyperparameters

The robustness of a decision tree model is highly dependent on the hyperparameters used to build the decision tree model in the first place. While a decision tree is fairly easy to understand, build, and interpret, it does have a number of hyperparameters that need to be defined during the training process.

What are Hyperparameters?

In machine learning, hyperparameters are parameters that you need to set before the learning processing can begin. They are only used during the training process and contribute significantly to the model training process. In contrast, model parameters are parameters derived as a result of the training process. Different model training algorithms require different hyperparameters, and understanding what these are and how to assign and optimize them is a big part of what the data scientists need to do.

The default parameters used in SAS Viya visual modeling are generally acceptable to build your initial model, but understanding and tuning your hyperparameters should be a critical part of your model building process. Figure 8.18 highlights the set of default hyperparameters used during our decision tree model training process.

Figure 8.18: Default Hyperparameter Values for Decision Tree

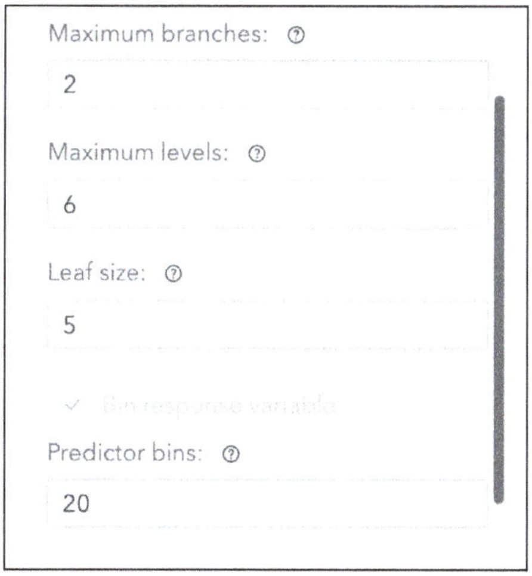

Maximum branches: ⓘ

2

Maximum levels: ⓘ

6

Leaf size: ⓘ

5

✓ Bin response variable

Predictor bins: ⓘ

20

As a citizen data scientist tasked with tuning and improving a machine learning model, one of the most difficult jobs is the assignment of these hyperparameters. The ability to assign and tune them effectively is highly dependent on experience, intuition, and understanding of the business problem. Fortunately, SAS Viya supports an autotuning capability, which makes it extremely easy for you to work out what the best hyperparameters should be. The hyperparameters autotuning function within the SAS Viya visual modeling environment will try to work out different hyperparameter combinations and optimize for the set that maximizes the performance of your model automatically. Figure 8.19 demonstrate the hyperparameter autotuning configuration window. It shows what hyperparameters will be optimized (maximum level and predictor bins in our example using a decision tree algorithm) as well as how long you would like to run the autotuning process for.

Figure 8.19: Configuring Hyperparameter Autotuning

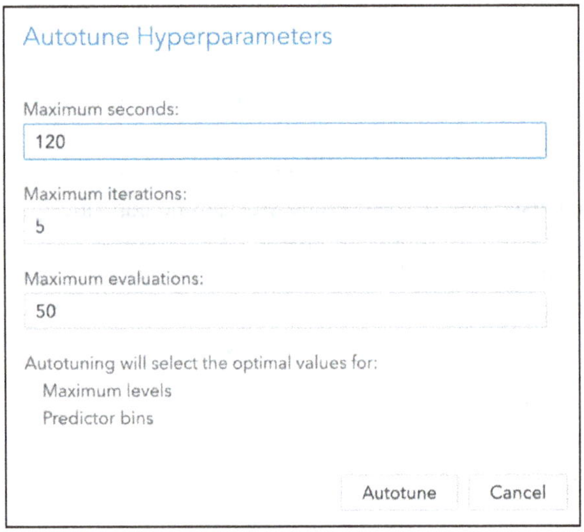

Autotune Hyperparameters

Maximum seconds:

120

Maximum iterations:

5

Maximum evaluations:

50

Autotuning will select the optimal values for:
Maximum levels
Predictor bins

Autotune   Cancel

When the autotuning is applied in our example, the resultant decision tree turns out to be a much improved version. The misclassification rate has dropped farther from 0.1209 to 0.0986. In addition, we have also improved the cumulative lift rate to 1.98 (from 1.56) at the 40th percentile as shown in the cumulative lift chart in Figure 8.20.

Figure 8.20: Improved Cumulative Lift Chart

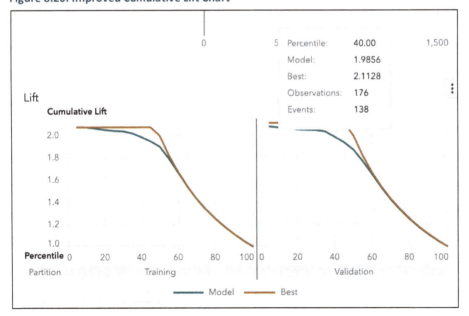

When we examine the decision tree window and the tree structure itself as shown in Figure 8.21, we can see that we have created a much more complex tree structure through the autotuning process, which can often be the case.

Figure 8.21: New Decision Tree Based on Autotuning

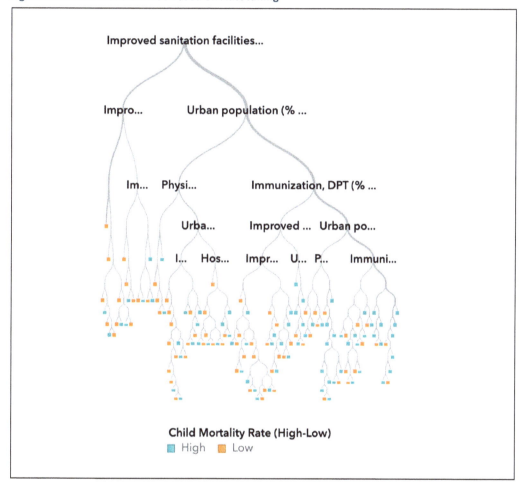

Tree Pruning

Very large decision trees such as the ones that we have produced from the autotuning process can be difficult to interpret and prone to overfitting. The process of tree pruning enables you to reduce the complexity of a decision tree and is often something that you need to do as part of the model tuning process.

The SAS Viya visual modeling environment supports the ability to prune a decision tree using manual or automated methods. The interactive mode enables you to manually grow or split the tree at the individual branch and gives you more control in how you want to prune and simplify a decision tree. Alternatively, you can also prune your decision tree automatically by setting specific model configuration options.

We will use the automated method to prune our decision tree by making the following model training option changes as shown in Figure 8.22:

- Uncheck **Prune with validation data**
- Move the **Pruning** dial to **80%**
- Uncheck **Reuse predictors**

Figure 8.22: Pruning the Tree Using Automated Options

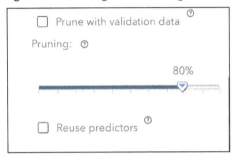

The Pruning dial controls how aggressively you would like to prune your tree and the Reuse predictor check box controls whether you would like to reuse your predictors. These are two of the most common controls that enable you to prune and simplify your tree automatically. The new tree created as a result of these changes is a much less complex tree as shown in Figure 8.23. It has fewer levels, fewer leaf nodes, and fewer splits, which make it much easier to interpret when compared with the previous tree.

Figure 8.23: Simpler Tree After Automated Pruning

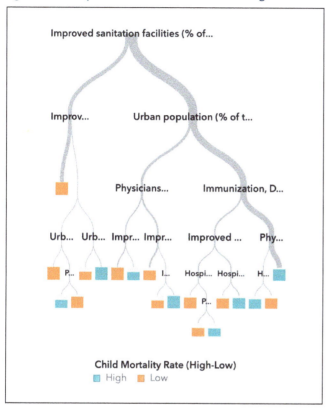

Child Mortality Rate (High-Low)

While the new decision tree is much simpler, the new misclassification rate has increased slightly from 0.0986 to 0.1163. In addition, the cumulative lift rate has also dropped slightly from 1.98 to 1.86. When we compare this decision tree with the decision tree before the pruning process, we can see that while both the misclassification rate and cumulative lift rate have suffered slightly, we have produced a much simpler decision tree model. It means that there is less chance of the model being overfit, and that it will also be a much simpler model to implement and put into production, both of which are important considerations when it comes to modeling.

It is worth noting that in addition to improving various model performance statistics, simplifying a model can also be an important aspect of tuning and improving your model. Sometimes these tuning steps produce performance results that are inferior relative to ones produced by a more complex tree as we saw in the previous example. Understanding the end goal of the modeling process and making the right compromise is a big part of the what the data scientist needs to do. Quite often, a model with slightly poorer performance but a much less complex structure is the more appropriate model to use.

When we revisit our original goals at the start of this chapter, we can now fully appreciate the ease and power of the SAS Viya visual modeling environment. Not only were we able to gain a much deeper understand of the child mortality rate and related factors, we were also able to train, tune, and create a robust and interpretable predictive model using a decision tree in an incredibly short amount of time. We have covered the basics in terms of building a predictive model using a decision tree, which is only one of many different techniques that a data scientist should be familiar with. We will be taking you through how you can build a similar predictive model using linear regression in the next chapter.

Chapter 9: Predictive Modeling Using Linear Regression

Overview

Linear regression is one of the oldest and most commonly used predictive modeling techniques. It quantifies the relationship between one or more predictor variables and a target variable and is commonly used to predict interval or continuous values.

Linear regression is considered the workhorse of statistics-based predictive modeling. The technique is mature, easy to understand and interpret, and widely accepted in highly regulated environments such as the banking and insurance industries. While it often does not produce the most accurate or robust predictive model relative to what can be produced using more modern machine learning techniques, its simplicity often makes it a great option to approximate a solution and form a baseline model when it comes to tackling regression problems.

In simple linear regression, the model consists of only one input variable (x) and one output variable (y). The relationship between the input and output variable is expressed as a mathematical equation of the form:

$$y = a + bx$$

Where "a" is commonly referred to as the Y-axis intercept, and "b" is referred to as the coefficient. The objective of the simple linear regression analysis is to identify a straight line that approximates the relationship between x and y in the form of the linear regression formula. Once a linear regression formula has been identified, not only can future target values be predicted, the parameters identified (a and b) can then also be used to understand the influence of the input variables.

In real-world applications, we often need to use multiple input predictor variables in a regression analysis and extend beyond using just one predictor variable. This is where linear regression (sometimes called multiple linear regression) comes to the rescue. Linear regression takes a similar approach but extends the ideas of simple linear regression by allowing multiple input variables. Just like simple linear regression, the parameters identified in a linear regression model allow us to interpret the model through the lens of the parameters identified. Compared with simple linear regression, multiple linear regression enables us to tackle more complex problems, build more accurate and robust predictive models, and at the same time retain similarly high degrees of interpretability.

The decision to use linear regression to build a predictive model should be taken carefully and be based on a number of important assumptions around the problem at hand:

- There should be a linear relationship between the input variables and the target variables. While this is fairly self-explanatory, in situations where you know the relationship between the input and output variables is definitely not linear, then you should consider using other techniques such as a decision tree to tackle regression problems.

- The input variables that you choose to use should not be highly correlated with each other. Linear regression can suffer from multicollinearity. In simple terms, multicollinearity describes situations where two or more predictor variables are highly correlated with each other. Multicollinearity is not desirable because it can make the parameters estimates (what we get through the model training process) very sensitive to minor changes in the model and make the model overfit. The result is that the parameter estimates are unstable, and a poorly fit model is created, so where possible, you should avoid using multiple input variables that are potentially highly correlated with each other.

What is Logistic Regression?

A close cousin of linear regression is the logistic regression modeling technique. While linear regression is used to predict continuous values, logistic regression uses a similar mathematical approach, but it can be used to predict discrete values and tackle classification problems.

Logistic regression typically requires the use of transformation functions against the input and output variables before they are used for the model training process. It is through this process of transforming the variables (log-based transform for example) that logistic regression becomes suitable to tackle binary or classification problems. The logistic regression model is supported as a dedicated modeling object in the SAS Viya visual modeling environment, and a number of transformation functions are built into the solution to make the process of building logistic regression models easier and more streamlined.

When we include multiple variables in the linear regression model, the analysis and model generated actually give us an estimate of the linear association between each input variable and the target variable while holding all other input variables constant (or "controlling for" these other factors). This unique characteristic of linear regression provides us with powerful insights around the importance and relevance of individual input variables, and that is why linear regression models are highly interpretable and can be used in highly regulated environments.

Model Building and Assessment

The linear regression object offered within the SAS Viya visual modeling environment supports multiple linear regression analysis where multiple input variables can be used to predict a single target variable. SAS Viya also includes detailed model fit and summary statistics that are useful for the modeler to understand and tune and improve the model.

We will use the linear regression model object to predict the child mortality rate of various countries. Instead of predicting whether a country will have high or low child mortality rate as we did in the previous chapter, we will treat it as a regression problem instead and try to predict the actual child mortality rate ("Mortality Rate, Under-5") in the following example.

Duplicating an Existing Predictive Model

Often, it can be beneficial to avoid creating a model from scratch and instead duplicate one from an existing model first. This is helpful if you are trying to solve the same types of problem (regression or classification) and would like to reuse the same set of input and output variables.

You can easily duplicate a model in the SAS Viya visual modeling environment by right-clicking on a model object and then selecting **Duplicate as** (see Figure 9.1). By duplicating the model, SAS Viya uses the same set of predictor and target variables and creates a new model using the new modeling technique selected. You can even duplicate the new model as a new page by right-clicking and then pressing the **Alt** key while selecting, which will give you an additional option to duplicate the selected model as a new page.

Figure 9.1: Duplicating an Existing Modeling Object

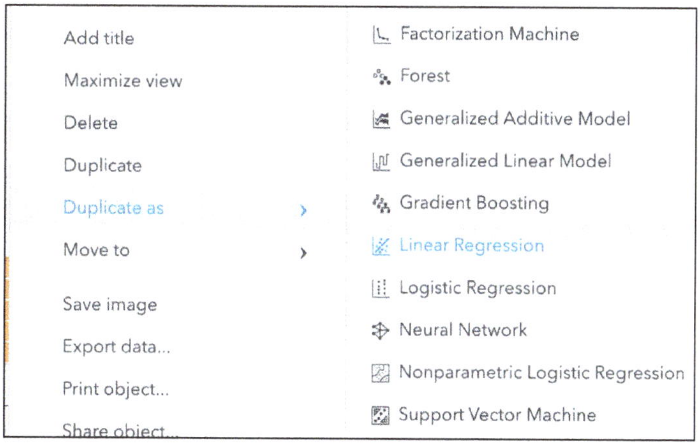

The SAS Viya linear regression object uses the least squares method to train and build a linear regression model. The least squares method creates a line of best fit by minimizing the residual sum of squares for every observation in the input data set. The residual sum of squares is the vertical distance between an actual observed value and the predicted value from the model (sometimes called line of best fit). The least squares method requires no assumptions about the distribution of the input data, which makes it more flexible and easier to use because you do not need to transform the input variables in order to use a linear regression technique.

Data Preparation

Similar to how we built the decision tree model in the previous chapter, we first need to partition our data into training and validation data sets in order to make sure that we do not produce a model that is overfit. Just like what we did in the previous chapter, we will be creating our new partition variable using the random sampling technique to create two data partitions – training and validation – with a 70/30 split.

As mentioned before, the linear regression model created using the least squares method works better when the variables are normally distributed but does not assume or require the variables that you use to be of a certain distribution. This means that we are able to quickly prototype linear regression models without applying any additional variable transformation before they can be used. And since our target variables are already in the correct format (Measure), we can start building our model without any addition data preparation steps.

Building the Linear Regression Model

We will only be using continuous effects (interval values) in our linear regression model example. We will start by using the same set of predictor variables that we used in the previous chapter:

- GDP per capita
- Health expenditure per capita
- Hospital beds (per 1000 people)
- Immunization, DPT (% of children ages 12-23 months)
- Improved sanitation facilities (% of population with access)
- Improved water source (% of population with access)

With child mortality rate ("Mortality Rate, Under-5") defined as our target variable and using all the default linear regression model configurations, we get our first linear regression model as shown in Figure 9.2.

Figure 9.2: Linear Regression Model

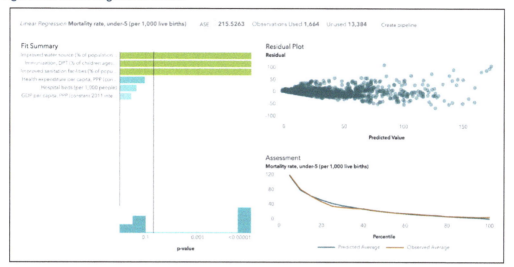

The top of the linear regression object highlights the target variable chosen (child mortality), the overall fit statistics (ASE of 215.5263), and the number of actual observations used by the model. Note that out of the total 13,384 observations, we ended up using only 1,664 observation in our model. This is due to the fact that any observations with missing values will be dropped when building a linear regression model, and we have a fairly sparse data set with lots of missing values in our training data set. A number of steps can be taken to increase the number of observations you can use to train the linear regression model, which is generally considered best practice. This includes removing variables with a large number of missing values or computing and filling in the missing values so that more data can be used during the model training process.

Overall Fit Statistics

SAS Viya supports a number of common model fit statistics when building linear regression models including AIC, R-Square, ASE, and SBC. These measures all help you understand the overall characteristics and fit of the model. For our model build, we will be focusing on the average squared error (ASE) value, which is calculated as the sum of squared errors divided by the number of observations. In other words, the ASE value highlights the average difference (error) between the actual value (child mortality rate in our case) and the predicted value across all the data points. ASE is a simple measure that is easy to understand and communicate and one that you want to minimize as you tune and improve your model to reduce the error or residual from each predicted value.

As shown in Figure 9.2, our model has resulted in an initial ASE of 215.52, meaning that on average, our model has produced predictions that are off from the actual value by 215.52. One of the goals of our model tuning steps is to reduce this value so that we can predict the child mortality rate with a higher degree of accuracy and produce a model with better fit.

Fit Summary Window

In additional to the overall model fit statistic, the model fit summary window (as shown in Figure 9.3) provides insights into the influence and significance of the input variables used.

Figure 9.3: Model Fit Summary Window

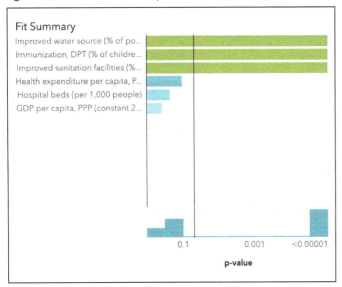

The fit summary window lists all the variables used in the model ranked by their p-values, which are used to determine the significance of the variables. The p-value is plotted on a reverse log scale, which means that the lower the p-value, the longer the bar and the more significant the variable is. When you position your pointer over the individual bars next to the variables, a pop-up window is shown with additional details including the actual p-value of each bar as well as the transformed p-value.

What Is a p-Value?

The p-value (probability value) is an important statistical measure and one of the best ways to test and validate if the results from an experiment are statistically significant. P-value is calculated using the statistical hypothesis testing framework. It can be explained as the probability that for a given statistical model the null hypothesis is true.

In our example, the null hypothesis is that variable X (rate of access to improved water source) does NOT contribute or affect the child mortality rate. The p-value represents the probability that our null hypothesis is true. In other words, the higher the p-value, the more likely our null hypothesis is true and the more likely variable X does not contribute or affect child mortality rate. Looking at it another way, a low p-value would indicate that our null hypothesis is not true, therefore indicating that the variable (rate of access to improved water) does contribute and affect the child mortality rate.

In our example, we can see that access to improved water source, immunization date, and access to improved sanitation facilitation all have long bars, meaning that they all have small p-values and are all important factors when it comes to predicting the child mortality rate using our linear regression model.

Residual Plot

The residual plot is one of the most important and useful assessment plots when it comes to building and analyzing a linear regression model. It shows the relationship between the predicted value of an observation and the residual of an observation. The residual of an observation is essentially the difference between the predicted target value and the actual target value.

$$Residual = Observed - Predicted$$

As shown in Figure 9.4, the SAS Viya visual modeling environment automatically produces a residual plot where the predictions made by the model are shown as the X axis, and the accuracy of the prediction is on the Y axis (shown as the residual value for that predicted value). The vertical distance between the actual predictions and the horizontal line at zero at any given predicted value shows the amount of residual value and how bad the prediction is for that value. Positive values for the residual (points above 0 on the Y axis) mean that the prediction was too low, and negative values (points below the 0 on the Y axis) mean that the prediction was too high; 0 means the prediction was exactly correct.

Figure 9.4: Residual Plot

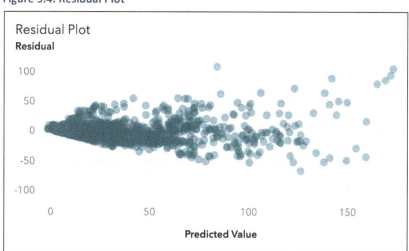

The overall pattern of a residual plot is very useful and can often reveal insights that can help you tune and improve the model. Furthermore, when you position your pointer over the individual residual points, a pop-up window is displayed (as shown in Figure 9.5) with additional information highlighting the predicted value, the residual value, and the associated input variable values. This is useful when you want to drill into individual predictions to better understand the input variables that might have contributed to an outlier or a large residual value.

Figure 9.5: Details Window from Residual Plot

Predicted Value:	173.12525109
Residual:	106.37474891
Health expenditure per capita, PPP (constant 2011 international $):	35.94455315
Hospital beds (per 1,000 people):	0.119999997
Immunization, DPT (% of children ages 12-23 months):	23
Improved sanitation facilities (% of population with access):	5.2
Improved water source (% of population with access):	38.7
GDP per capita, PPP (constant 2011 international $):	784.3531743

In this case, we can see that we have a predicted value of 173.125 for child mortality rate with a relatively large residual value of 106.37. In other words, our prediction is off the actual value by 106.37 for this set of variables. This is something that we potentially need to investigate further in order to improve our model

A balanced and stable model should obviously only contain points with low residual values, but the overall residual plot generally should also have the following characteristics:

- It should be symmetrically distributed, tending to cluster toward the middle of the plot.
- It should be clustered around the horizontal axis where y=0.
- In general, there should not be any clear patterns. Obvious patterns in the residual plot indicate that the model might not fit the data well

In our case, we can see that our residual plot exhibits the following characteristics:

- As the predicted value increases (toward the right-hand side), the residual seems to increase. This is especially the case when the predicted value is below 50.
- It contains points with relatively large residual values (more than 100).

The presence of points with large residual values means that we potentially have outliers that are skewing and effecting the overall fit of our model. In addition, when we can detect a clear pattern or trends in our residual plot, then it often means that our model has room for improvement where we can capture the pattern in our model. These are things that we potentially need to address during the model tuning process.

Another useful way to examine the residuals plot is to analyze it as a histogram. As we discussed in Chapter 3, a histogram enables you to view measures (residuals in this case) as a histogram distribution. The SAS Viya visual modeling environment enables you to view the residual plot as a histogram by right-clicking on the residual plot and selecting **Use histogram**. The same residual plot values from our linear regression model can now be seen as a histogram as shown in Figure 9.6.

Figure 9.6: Residual Plot Histogram

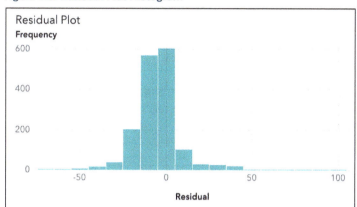

Ideally, you would like to see a normal distribution centered around zero, which would indicate a model that fits well and is stable. Our residual plot histogram shows a histogram that is skewed slight to the right and not really centered around zero. This would be another area that we would want to improve upon as we tune the model in later steps.

Details Table

When you expand the linear regression object, a details table becomes available at the bottom of the main canvas. The details table provides additional metrics and details associated with the linear regression model. Two of the most useful ones are the fit statistics table and parameter estimates table.

The fit statistics table as shown in Figure 9.7 includes all the model fit statistics in a single table. This is useful because it enables you to assess other important model fit statistics such as R-Square and Mean Square Error all at the same time. Above and beyond the ASE value that we have already discussed earlier, the R-Square value measures the strength of the relationship between our model and the target variable. In our case, we can see that the R-Square value for our model is 0.80099, which indicates that our model can explain approximately 80% of the variability of the target variable (child mortality rate).

Figure 9.7: Fit Statistics Details Table

Dimensions	Overall ANOVA	Fit Statistics	Parameter Estimates	Type III Test	Assessment	Assessment Statistics

Statistic	Value
Mean Square Error	216.4368
Root MSE	14.71179
F Value for Model	1111.539
Pr > F	<0.00001
R-Square	0.80099

The parameter estimates table as shown in Figure 9.8 displays the parameter estimates, standard error, t-value, and p-value, all of which help you in assessing the significance and impact of input variables for the model that you have built.

Figure 9.8: Parameter Estimates Details Table

Dimensions	Overall ANOVA	Fit Statistics	Parameter Estimates	Type III Test	Assessment	Assessment Statistics

Parameter	Estimate	Standard Error	t Value	Pr > |t| ▲
Intercept	236.4803	3.682314	64.22058	<0.00010
Improved water source (% of population with access)	-1.1357	0.053435	-21.2538	<0.00010
Immunization, DPT (% of children ages 12-23 months)	-0.73923	0.043284	-17.0787	<0.00010
Improved sanitation facilities (% of population with access)	-0.45564	0.031384	-14.5183	<0.00010
Health expenditure per capita, PPP (constant 2011 international $)	-0.00075	0.000461	-1.62049	0.10532

The table can be sorted by clicking on the individual columns (as we have done in Figure 9.8 against the last p-value column). Similar to what saw from the fit summary window, we can see that access to improved water sources, rate of immunization, and access to improved sanitation facilities are the three most significant variables (low p-value).

In addition, the parameter estimates table also provide us with the parameter estimates values (or coefficients) for the individual input variables. These parameter estimates describe the relationship between the predictor variable and the target variable in the form of a simple multiplier. The sign of each parameter estimate indicates the direction of these relationships. These parameter estimates values represent the relative change in the target variable given a one-unit change in the predictor variable. In our case, we can see that rate of access to improved water sources has a negative parameter estimate value (-1.1357) and therefore a negative relationship with the child mortality rate measure. This means that as we improve or increase the rate of access to improved water sources, child mortality rate should decrease for that country. Furthermore, for each increase in rate of access to improved water source, we should see 1.1357 times decrease in the child mortality rate value.

Interpreting the Model

As discussed earlier in this chapter, the output of a linear regression model is essentially a mathematical formula. By combining and leveraging all the parameter estimates values, we can come up with the mathematical formula underpinning the linear regression model that we have created. Not only does this formula provide us with a complete view of how the various predictor variables contribute to the target variable, it also allow us to interpret and communicate the model with others in a simple way, which is an important aspect of what a citizen data scientist needs to do.

If we take the top 3 most significant variables (as determined by the p-value) from the parameter estimates table, we can describe our linear regression model using the following mathematical formula:

$$Y = (-1.1357 \times A) + (-0.73923 \times B) + (-0.45564 \times C) + 236.4803$$

- Y = Child mortality rate
- A = Improved water source (% of population with access)
- B = Immunization, DPT (% of children ages 12-23 months)
- C = Improved sanitation facilities (% of population with access)
- Intercept = 236.4803

We can then interpret the model and make the following observations:

- All three predictor variables have a negative relationship with the target variable. As we increase all the predictor variables (rate of access to improved water sources and so on), the target variable would decrease.

- For each increase in rate of access to improved water source, we should see a 1.1357 times decrease in the child mortality rate value.

- For each increase in rate of immunization, we should see a 0.73923 times decrease in the child mortality rate value.

- For each increase in rate of access to improved sanitation facilities, we should see a 0.45564 times decrease in the child mortality rate value.

With a simple-to-use, drag-and-drop interface, the SAS Viya visual modeling environment makes it extremely easy to build a basic linear regression model in a short period of time. The fit summary window, residual plot, and details table all provide valuable insights around the significance and impact of the input variables used. As a discovery tool, it can shed unique insights that traditional visualization techniques are unable to provide. But the visual modeling capability of SAS Viya also extends further into model tuning, which we will now examine.

Tuning and Improving the Model

With the initial model performance and fit statistic baseline established, we now turn our attention to tuning and improving the linear regression model. Similar to the approach taken when building the decision tree model in the previous chapter, we will be leveraging the various model fit statistics and highlighting some of the most common steps that you can take in order to improve a linear regression model.

Let's start by clarifying the key objectives that we want to achieve in terms of improving our model. In this case, we want to see whether we can make incremental improvement to our model by focusing our efforts on the following:

- Improve the overall model performance by decreasing the ASE
- Reduce and eliminate any extreme outliers and patterns in the residual plot
- Reduce the overall complexity of the model

Adjust Input Variables

From the initial model fit summary window, we saw that the input variable "GDP per capita" plays an insignificant role (high p-value of 0.3561) in influencing child mortality within the model that we have built. By removing this input variable, we are able to make marginal improvements to the model in the following ways:

- Reduce the overall complexity of the model
- Improve the number of observations that can be used for the model training process (from 1664 to 1702)

Removing the "GDP per capita" variable has the positive side effect of giving us more training data, which is always good. Records that we had to omit previously since the "GDP per capita" value is missing can now be used in the training process (if all other variables are not missing). With just the small increase in the training data set, we saw a marginal improvement in terms of the overall ASE (from 215.5263 to 214.8443).

Removing Outliers

The residual plot from the original model built highlighted a number of predictions with large errors or residual values. This is especially the case when the predicted value was above 80. These predictions with high positive residual values meant that we have a number of data points in our training data set where the actual value was much higher than the predicted value. These outliers in the residual plot are generally considered undesirable, and they indicate that our model will be skewed and might not be stable or produce a good overall fit.

These outliers normally would warrant further inspections to work out why they exist and can be dealt with in a number of ways. There might be legitimate reasons as to why they exist (specific countries with high child mortality rates for example) or they might be due to data collection error. Based on this further inspection, you can improve the model and produce a more desirable residual plot by either transforming the variables or removing these outliers all together.

In our case, we will assume that these points with large residual values represent extreme cases due to collection error, and we will remove them in order to produce a model that has a better overall fit. The SAS Viya visual modeling environment enables you to examine and remove data points directly from the residual plot itself. To examine one or more points from the residual plot, first select those points in the residual plot itself (as shown in Figure 9.9).

Figure 9.9: Selecting Multiple Points on the Residual Plot

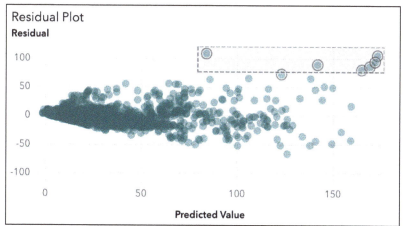

With the points that you want to examine further selected, right-click and select **Show selected**. This opens a new list table window containing only the selected points as shown in Figure 9.10. Examining these points further gives you additional insights as to what might be the causes of the high residual values and what to do about them.

Figure 9.10: Showing Selected Points from Residual Plot

Show selected ×

Predicted Value	Residual	Mortality rate, under...	Health expenditure per c...	Hospital beds (per 1,000 pe
172.07262929	107.42737071	279.5	35.94455315	0.119999997
164.04647018	81.753529818	245.8	41.28029933	0.119999997
168.11770039	87.782299605	255.9	37.41266761	0.180000007
170.79532462	96.304675376	267.1	35.04809299	0.119999997
136.90123072	66.998769276	203.9	157.521163	0.8
141.42859796	90.271402044	231.7	60.2521898	0.239999995
123.24368581	72.956314188	196.2	116.9760688	0.4

Close

Once you are satisfied with having the selected points removed, right-click over the selected points again and select **New filter from selection** and then select **Exclude selection** (as shown in Figure 9.11). A new linear regression model will now be built without the selected outlier points.

Figure 9.11: Excluding Points from the Residual Plot

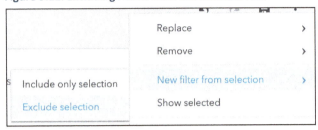

We can now see a new residual plot (as shown in Figure 9.12) with no extreme values (higher than 70). More importantly, we have now produced a model with a much-improved overall AES value (from 214.8443 to 163.2394).

Figure 9.12: New Residual Plot with Outliers Removed

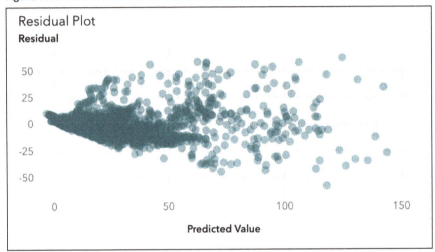

Switching the residual plot to histogram (as shown in Figure 9.13) tells a similar story. Compared to the original residual plot histogram, we can see that this new histogram is closer to a normal distribution with less right skewness, which is what we would like to see in a good, stable model.

Figure 9.13: New Residual Plot Histogram with Outliers Removed

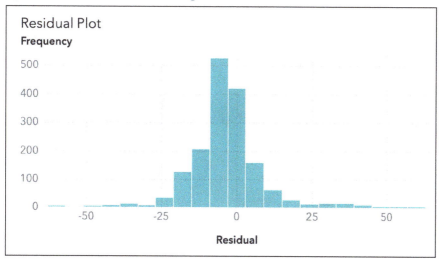

Adding Group-By Variables

Group-By is a specific modeling technique that enables you to build multiple models grouped by a specific variable. It is an extremely useful technique, especially when you have natural groupings in your data set that have unique characteristics in relation to the event that you are trying to predict. Building multiple models means that you are able to capture the relationship between the input and output variables more precisely within each of the groups.

In our case, it is likely that there might be region-specific factors that influence the child mortality rate. As such we will try to group our data set by the region variable and build multiple linear regression models in order to produce a model with better overall fit. You can easily create multiple linear regression models using the "Region" variable by adding and defining it as a **Group by** variable in the **Roles** configuration section as shown in Figure 9.14.

Figure 9.14: Adding Region as a Group-By Variable

With the new group by variable added, the linear regression model object refreshes with a new linear regression object as shown in Figure 9.15.

Figure 9.15: New Linear Regression with Group-By Variable

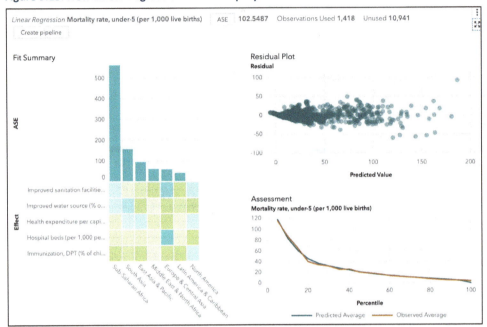

The first thing that we can see and perhaps the most significant change is that we have made further improvements in terms of the overall ASE value (from 163.2394 to 102.5484). This is largely due to the fact that we have now created seven linear regression models all individually calibrated using data from specific regions. Instead of trying to fit and produce one model with all the data and input variables using a single linear regression formula, each of the regions now has a specific linear regression model with parameter estimates finely tuned to only data from that region.

The performance characteristics (ASE values as bar chart) of these models and the variable effects can now be seen in the new fit summary window, which is broken down by the regions. When you position your pointer over the model fit statistics bar chart, a pop-up window displays the number of observations used and the ASE value for the model built using that subset of data as shown in Figure 9.16.

In our new model, we can see that while the model built using data from the sub-Saharan African region (first bar in the fit summary window) have produced a relatively poor model with a high ASE value (562), models built using data from all the other regions seems to have all produced relatively good models with low ASE values. For example, the model built using data from East Asia & Pacific resulted in an ASE value of 96.71 (as shown in Figure 9.16), which is the lowest ASE value that we have achieved thus far.

Figure 9.16: Selecting a Model Grouped by Region

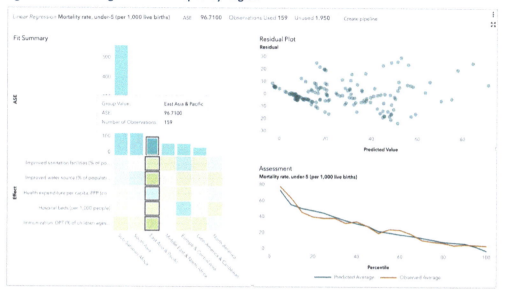

The whole linear regression model object automatically refreshes when we select a model in the fit summary window on the left-hand side (as shown in Figure 9.16). The updated view includes a filtered view where the overall fit statistics (ASE), residual plot, and assessment plot only reflect performance characteristics from the model that you have selected (East Asia & Pacific in this case). This filtered view provides additional insights regarding each of the individual models that you have created, which enables you to inspect and tune them individually. In this case, we can see that the model trained using data from the East Asia and Pacific region has a relatively low ASE value and also a residual plot that does not exhibit obvious patterns or include any extreme outliers, which is what we would like to see.

Using a combination of different techniques including variable selection, outlier removal, and Group-By, we managed to improve our linear regression model in a relatively short period of time. Similar to what we saw when building the decision tree model, we also managed to uncover valuable insights about the target and predictor variables along the way. We developed a powerful narrative about the business problem by interpreting what the model and data revealed to us.

As we have seen in this and the previous chapter, the SAS Viya visual modeling environment makes it extremely easy to build and tune machine learning models. Not only is the visual interface of SAS Viya great to discover valuable insights during the modeling process, the various model fit statistics and performance chart also make it extremely easy to tune and improve the model on the fly.

Perhaps the best thing about the visual modeling environment is that once a good model has been built, you can then easily share and communicate what you have built to anyone via the rich, visual interface. The ability to effectively communicate findings to key stakeholders regardless of their personas is critical, and that is where the visual modeling capability of SAS Viya really shines.

Chapter 10: Bring It All Together

Combine and Experiment

"The whole is greater than the sum of its parts."

– Aristotle

Visualizing data, analyzing trends, and predicting outcomes in the real world can be a messy and complicated affair. Often you want to understand the complex interactions between many different variables and factors at the same time. You want to apply filters from different perspectives, highlight a sub-segment as you navigate through different charts, and understand the influencing factors for an event in different geographical regions on the fly. This is where combining different visualizations and charts in a dynamic and interactive way can really help. The ability to combine and link different visualization charts and objects on the same page, both traditional statistical charts as well as visual modeling objects, is where the power of smart discovery and SAS Viya really get unleashed. When you combined objects intelligently and in a thoughtful way, amazing insights can be discovered, and the whole is definitely greater than the sum of its parts.

When it comes to combining different charts and using advanced interactions, it is important to make sure that you can walk before you start running. Start by mastering the basics of the various visualization objects and learning the foundational technique and interactions supported by the SAS Viya visual modeling environment (by reading the first nine chapters of this book!). Once you have mastered the basics, combining charts and visualizations not only enables you to ask questions with more depth, it also enables you to apply your business domain knowledge and test your hypothesis in interesting and unique ways.

Combining Visualizations

While each of the SAS Viya visualization objects are powerful and dynamic in their own right, what makes them even more useful is the fact that they are not black boxes that exist in isolation. Objects can be combined and linked in order to interact with each other interactively. You can combine and enable interactions using different visualization objects in SAS Viya including basic chart objects as well as advanced visual modeling objects such as the decision tree and linear regression objects. They can be combined and linked using filter or linked selection actions, and these interactions can often be bi-directional as well. There really is no technical limitation to how you can combine all the different visualization objects together. The only limiting factors are the types of problems that you are trying to solve and your own imagination.

For example, we can build on the final decision tree model that we developed from Chapter 8 by adding a bar chart to the same page and linking the two objects together using a filter action as shown in Figure 10.1.

Figure 10.1: Linking the Decision Tree and Bar Chart Using a Filter Action

We end up with a page (as shown in Figure 10.2) where on the left-hand side we can see regions and average life expectancy displayed as a bar chart, and on the right-hand side we have a predictive decision tree model predicting child mortality rate. With the two objects linked using a one-directional filter action, we can now filter the decision tree on the right-hand side by simply clicking on individual regions on the left-hand side.

Figure 10.2: A Page with a Linked Bar Chart and Decision Tree Model

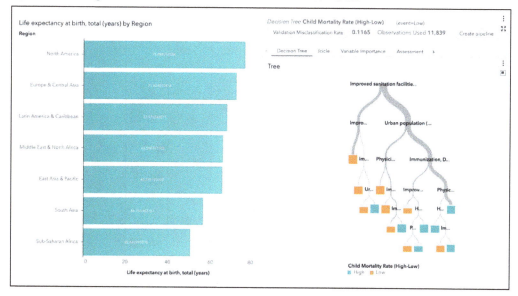

For example, from the bar char on the left-hand side, we can see that countries in the South Asia region have relatively low average life expectancy (56.35), and we want to drill in further to better understand the various influencing factors. Simply clicking on the South Asia bar on the left side will filter the data used to train the decision tree model on the fly, and you end up with a new decision tree on the right-hand side immediately with only data from countries in the South Asia region as shown in Figure 10.3. In this case, we can see that child immunization rate is a key factor when it comes to influencing child mortality (top of the tree) in the South Asia region. This factor is perhaps an important consideration as we think about how to raise the overall average life expectancy for people living the South Asia region.

Figure 10.3: Interacting and Applying Filter to the Visualization Objects

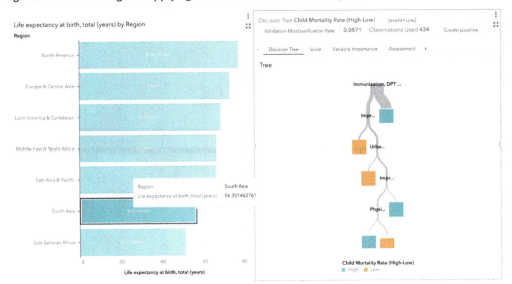

This type of fluid, dynamic interaction supports a natural flow of analysis and questioning. It enables you to test theories and hypotheses on the fly with minimum friction and can often lead to valuable insights and discoveries involving multiple factors.

Experimenting

Perhaps the most important thing that you can do in order to advance your smart data discovery skills is to simply try, experiment, and learn from doing so. The nature of experimentation means that sometimes you will find incredible insights as you try different charts and ways of combining them. Sometimes, experimenting with different chart combinations might lead you nowhere, and that is perfectly okay.

The dynamic, interactive nature of SAS Viya means that there is minimal friction and wait time as you experiment with new ideas. The cost of trying and failing is very small, but the potential upside and reward could be significant. After all, you are only one undo (or three!) click from where you were previously.

No matter what your specific problem is, follow your curiosity and intuition and get outside of your comfort zone. Here are some things to try as you start on your journey of experimentation and discovery using SAS Viya:

- Add a new data source or join a new data set to try new variables
- Change the type of chart that you use to visualize the same data set in a different way
- Try combining different charts and visual models together
- Change the type of interactions (linked selection, filters)

Share and Communicate

It is well-documented that analytics is very much a team sport today. Not only is it critically important to work and collaborate with other analysts and data scientists throughout the data discovery process, keeping important business stakeholders abreast of your progress and findings during and after the data discovery process can also make or break your analytics projects or initiatives. Often, the ability to share and communicate your findings effectively can be more important than the data discovery and data science skills themselves. This is why you should invest the time and effort to focus on the most effective way to share and communicate the insights and findings from your data discovery process.

Be a Storyteller
Figures and numbers represent facts, but narratives and stories convey emotion and build connections. It is said that people often base their decisions on emotion and use facts to back up their decisions. This is where the ability to use data and figures to tell a compelling story is the most powerful way to share ideas and inspire actions. How you present your findings in terms of the types of figures and charts presented, the communication platform and tools used to share your insights, and the narrative that you use to weave it all together can often mean the difference between a canceled project or gaining key stakeholders' support.

One of the key benefits of data visualization is the ease of sharing and communicating the charts and models that you have built to potentially any type of audience. Using a tool such as SAS Viya not only enables you to share your findings via the most effective mechanism quickly, you can also take advantage of the unique sharing platform and mechanisms supported to tell a compelling story to the right people via the right means at the right time.

Report Sharing

SAS Viya enables you to share your findings and reports in a number of different ways. The flexibility offered means that you can share your reports based how your intended audience prefers to ingest information and insights.

The most basic (and some would say most primitive!) way of sharing your insights via SAS Viya is to simply share your report as a PDF document. The PDF can be printed and enables you to share your reports without the Internet or a computing device. This method is often suitable in meetings where having printed copies of reports (that can be drawn or written on) is the most effective way to convey ideas and collaborate. The obvious downside of having a static PDF or printed report is that it is not dynamic. You lose the ability to filter and slice your data on the fly. The inability to drive live interactions and filter your data on the fly severely constrains your ability to tell a powerful and compelling story, so think carefully before you generate PDF reports or print your reports.

Alternatively, you can share your reports and findings via email or as a link to the report directly. Viewing the reports and visualizations via the browser provides you and the viewers with full access to the dynamic interactions supported by SAS Viya. It is often the simplest and most effective means to share and collaborate around your findings. Full collaborative commenting is also supported via the browser interface and is a great way to discuss specific findings on the report asynchronously.

> ### Commenting at Multiple Levels
>
> When you need to comment or collaborate with others around a report, it is often useful to be very precise and exact as to what you are commenting about. The SAS Viya visual environment enables you to comment on objects at different levels. You can see this in the example shown in Figure 10.4 where you have the option of submitting a comment on either the whole report ("Decision Tree – Combined"), the bar chart only ("Bar – Region 3"), or the decision tree object only ("Decision Tree"). In our case, we are submitting a comment on just the decision tree object only for better clarity and context.

Figure 10.4: Submitting Comments at Multiple Levels

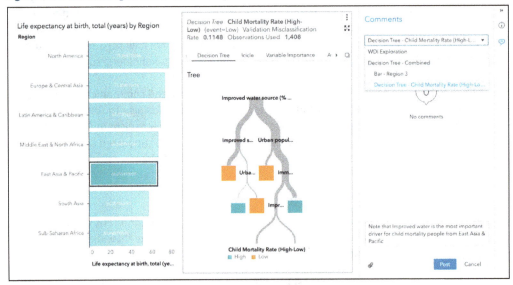

Mobile App

In an age where large amounts of information consumption are done on mobile devices on the go, the ability to share and collaborate via mobile devices is critical. SAS Viya supports the sharing and consumptions of reports and visualizations via dedicated mobile apps. There are mobile apps that are compatible with all the leading mobile device platforms including iOS, Android, and Microsoft Windows. These apps are designed to take advantage of the unique capabilities of each platform and are an effective way to share insights and collaborate with key stakeholders.

The SAS Visual Analytics mobile app on iOS showcases what is possible when you leverage mobile platforms for insight sharing and collaboration. The app enables you to interact with the reports and visualizations via a familiar look and feel (what you would see in a web browser) but takes advantage of the touch-based user interface offered on iPhone and iPads.

Figure 10.5 shows a typical screen of how a pre-built report would look like on an iPad using the SAS Visual Analytics app. Note that the charts are formatted specifically for the mobile device (iPad in this case), and you can interact with the report via the full range of interactions built into the report. Device-based push notifications can also be configured to notify the users when data within a report has been refreshed.

Figure 10.5: SAS Visual Analytics Mobile App

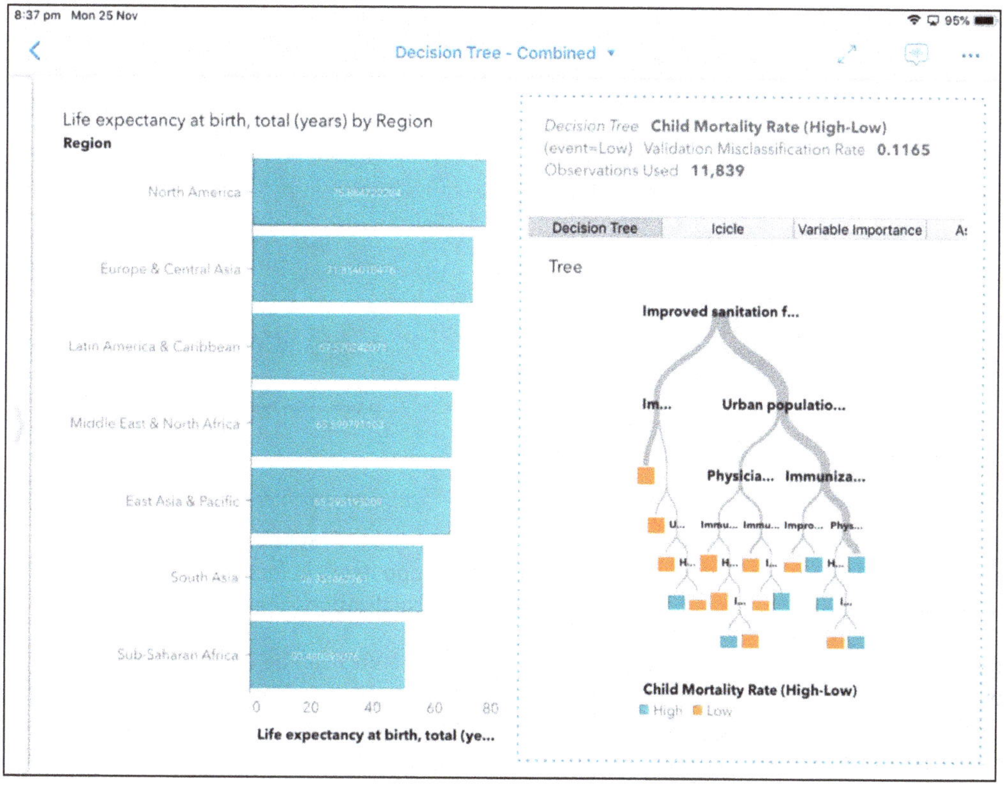

As you touch and select the bar chart on the left-hand side as shown in Figure 10.6, a pop-up window displays what you have selected, and the decision tree model is filtered and refreshed using the subset of data that you have selected. Essentially this is the exact same behavior that you would expect on a web browser, just optimized for the iPad screen and touch-based user interface.

Figure 10.6: Interaction Within the SAS Visual Analytics Mobile App

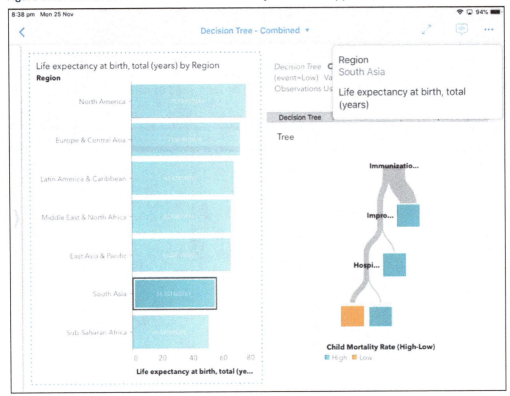

Not only does the app support all the standard interactions, it also supports interactive commenting, sharing, and annotation directly within the app itself. In fact, the only thing you cannot do on the app is edit and change the report and visualizations themselves.

All of the reports and visualization objects supported by SAS Viya can be viewed via mobile devices without any additional effort, which makes this an easy and effective way to share and communicate insights on the go. The interactive nature of these apps means that it is also a great way to collaborate with someone as you test ideas and hypotheses together. When it comes to being a great storyteller, you should always consider using the mobile app as a way to convey your findings, especially when you are away from a desktop PC and in a one-on-one situation.

Production and Deployment

A valuable and insightful finding made through the smart data discovery process can often be an opportunity to improve an existing business process and deliver additional business value. This means using the insight and extending it into a more advanced and robust predictive model that can then be embedded into a business process. This could be an internal or external customer-facing process that can potentially be used to improve and automate processes across multiple business functions.

For example, if you found through your data discovery process that customers with certain demographics tend to buy certain types of products together, it then becomes an opportunity to operationalize that finding by building a recommendation model that can be embedded into your online shopping process to improve the overall customer experience. Another example might be related to patterns that you discovered whereby certain types of hospital patients seems to have a high probability of re-admission. You can then build a predictive model and use this model to improve how hospitals review patients during the exit and follow-up process in order to reduce the chances of patients having to be re-admitted, which could benefit both the hospital and the patient.

Building and deploying a scalable and robust model typically involves multiple steps, a high degree of rigor, and input from multiple teams. It typically involves the collaboration of data scientists, data engineers, web developers, and the IT operations team. At a high level, the development and deployment of predictive models involve the followings key steps:

- Data engineering
- Pipeline model development
- Registering and management of models
- Model publishing and deployment
- Ongoing monitoring of models

SAS Viya was built to support the complete analytics life cycle and tackle these requirements in a seamless way. Using integrated modules within the SAS Viya platform such as Visual Data Mining and Machine Learning, Model Manager, and Intelligent Decisioning, teams can accelerate through the analytics life cycle and get to model deployment in a controlled and governed fashion.

Each of these analytics steps and SAS Viya solution components is really a book in its own right, but we will try to highlight some of the key downstream steps and requirements that you should be aware of in order to collaborate better with others when deploying models into production.

Pipeline Modeling

Pipeline modeling is a more comprehensive and robust way to build complex predictive models. It offers more flexibility and advanced functionalities that make it easier to build, test, compare, and improve models in a GUI-driven, interactive environment. Pipeline model development is supported via SAS Model Studio, which is a code-less model development interface offered as part of SAS Visual Data Mining and Machine Learning.

The integrated nature of SAS Viya means that you do not even have to build a pipeline model from scratch as you can turn an existing visual modeling object into a pipeline model. Once you have developed a visual modeling object using Visual Analytics (as we have done in Chapter 8 and Chapter 9), you can easily convert the object into a pipeline model in SAS Model Studio by simply right-clicking on the object and select **Create pipeline** as shown in Figure 10.7. The ease of transition between the SAS Viya components significantly reduces the friction between different modeling steps.

Figure 10.7: Creating a Model Pipeline from a Visual Modeling Object

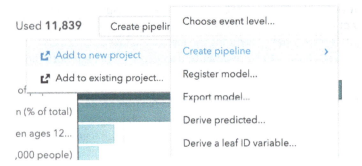

You can use SAS Model Studio to create a new model pipeline or edit an existing model pipeline using a drag-and-drop interface as shown in Figure 10.8. The objects on the left-hand side represent pre-built, common tasks that you can include in a modeling pipeline (such as feature selection and compute missing values). You then use the pipeline structure in the main canvas to link and sequence the nodes together as needed in order to prepare the data, train the model, and compare different models using different modeling techniques.

Figure 10.8: Developing Pipeline Model Using SAS Model Studio

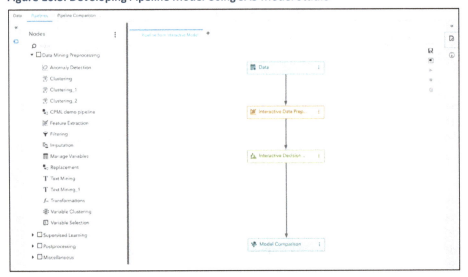

Pipeline-based modeling is not the only way to build an advanced model using SAS Viya. You can also do it programmatically. It is, however, an intuitive and rapid way to build and test complex models.

Model Management and Deployment

Model deployment is ultimately where you put the model into production and leverage it to score new data and make new predictions. This is often called the last mile, and it is where most data science teams struggle. SAS Viya offers comprehensive model management and model deployment support by centralizing and automating key steps associated with this last mile.

Once a robust model has been built (either via pipeline or programmatically using SAS Viya), it can then be registered into a central repository so that it can be tracked and accessed easily by multiple teams. More importantly, SAS Viya then simplifies and automates the process of publishing and deploying the model regardless of the publishing destination. SAS Viya models can be deployed as a batch process, a real-time scoring API, into a Hadoop environment, or embedded into a business decisioning process. The deployed model can then be monitored for performance decay and retrained or rebuilt as needed.

Model Management and ModelOps

There is a growing recognition that models are valuable corporate assets that need to be managed and governed in a robust and scalable way. Just like software source code or company transaction data, models used throughout an organization need to be tracked and managed centrally to drive better usage and manage risk.

ModelOps practices extend the concept of model management into the realm of operations and streamlines the requirements associated with model deployment and ongoing monitoring. There are strong parallels between ModelOps and DevOps, and both can work hand in hand to improve the overall effectiveness of IT and data science teams.

Although the journey of data science and machine learning often starts with smart data discovery, it can lead to other important downstream processes and activities as we have just highlighted. The value of an analytical insight multiplies as it is further refined and deployed into business operations, which is why you should be familiar with the complete analytics life cycle in order to collaborate with other stakeholders and deliver ongoing value for your organization.

Where to Go from Here

"Life is a journey, not a destination."

– Ralph Waldo Emerson

To say that this book has only scratched the surface of what is possible via smart data discovery and SAS Viya would be an understatement. The goal of this book is really to show you the possibilities by covering the fundamental concepts and highlighting key capabilities of SAS Viya. The sky really is the limit when you bring it all together with a little bit of curiosity and imagination. I hope I have achieved that goal and showed you the potential power and value of using SAS Viya for smart data discovery and started you on what will be an ongoing journey.

It is now up to you to not only build on the foundational knowledge gained in this book but apply it to your specific context and situation. It is through the process of application that you will likely find gaps in your knowledge. These gaps will then naturally lead you down the path of further learning and discovery. I have highlighted a number of important elements of smart discovery in this book, but here a few areas that I recommend you explore further as you advance your smart data discovery journey:

- Use more advanced data transformation techniques to build richer data sets and build better models

- Learn and leverage more advanced visual modeling objects and techniques (such as logistic regression and gradient boosting) to solve more complex predictions problems

- Gain a deeper understanding of downstream, more advanced model building and deployment techniques and steps

All of these are supported by the SAS Viya platform, but they all require knowledge and skills beyond what is covered just in this book. Learning more about smart data discovery has given you a great foundation to build on in order to advance your knowledge in each of these areas.

Ongoing Learning and Valuable Resources

The key to being successful and staying relevant in analytics and data science is not really about what you know today but being committed to be a lifelong learner. Technology and data science techniques are changing and evolving on a daily basis, and what you know now is likely to be outdated in the not too distant future.

As such, being humble and keeping yourself up-to-date by reading books such as this one and going to online or physical training courses on a regular basis are critically important in keeping your skills relevant and useful.

There are a number of helpful, free online resources that can help you expand your knowledge going forward:

- Official SAS product documentation (support.sas.com): The official SAS support and documentation website is the official knowledge repository when it comes to gaining in-depth understanding of relevant product features and functionalities. It is a great resource to leverage when you want to gain a deeper understanding of how to get more out of the SAS Viya platform.

- SAS Global Forum papers: SAS Global Forum papers are submitted by SAS product managers and experienced users from around the world and are a great source to learn more about latest trends and best practices. They tend to be very practical and focused on application and uses cases, which makes them a great resource for people who are already familiar with the basics (that would be people like you who have read this book!).

- SAS YouTube Channel: There is nothing better than watching someone do it when it comes to learning how to do something. The SAS YouTube channel has product demonstration videos that showcase the latest capabilities of SAS Viya and how they can be applied to solve specific business problems. It is a great resource for those of you who learn best by watching and following along.

Data can tell amazing stories, and it has the potential to transform every single organization in the world. By leveraging your curiosity and staying committed to the process of ongoing learning, you can truly make a difference in your organization, your community, and the world. But it is ultimately up to you to make that happen. It is about being humble, staying curious, and having fun while doing so.

I hope this is the start of a wonderful journey as you experience the wonder and power of data. It is also my hope that you truly make a difference in whatever role or industry you are in. The power to know is truly in your hands now.

Ready to take your SAS® and JMP®skills up a notch?

Be among the first to know about new books,
special events, and exclusive discounts.
support.sas.com/newbooks

Share your expertise. Write a book with SAS.
support.sas.com/publish

sas.com/books
for additional books and resources.

THE POWER TO KNOW®

www.ingramcontent.com/pod-product-compliance
Lightning Source LLC
Chambersburg PA
CBHW080530060326
40690CB00022B/5090